Creating with Mobile Media

"*Creating with Mobile Media* is a book that encapsulates the key paradigms of creative practice in a mobile world in a thought provoking, academically grounded and joyful approach to making personal, sharable art. Marsha Berry brings her deep insight and unique ability to cut through the jargon and situate creative practice in a context that can be understood by creative practitioners and theorists from fields including new media, cultural studies, literary studies, creative writing, screen studies and filmmaking. By sharing and interrogating her own creative practice via an accessible and deeply considered digital ethnological lens, Berry strikes new ground in the theorising of creative practice research in a mobile world. The smartphone as we all know is a vital part of everyday life in the 21st-century. Drawing on the image and practice of the wayfarer and "'being there" and reflecting on her own mobile art in video, photography and poetry, Berry's work stands as a key example of the seamless interrogation of creative practice in the academy. In *Creating with Mobile Media*, Berry brings fresh insight into the importance of the mobile phone as not only a means of communication but also a way of expressing ourselves, of thinking about and of making and sharing our art in an ever connected world".

—Margaret McVeigh, *Senior Lecturer and Head of Screenwriting and Contextual Studies, Griffith Film School, Griffith University, Australia*

"Marsha Berry writes on the cusp of anthropological investigation, poesis, and art praxis. Her writing is a self-aware and honest transaction with her reader. She does not want to overwhelm us with theoretical muscle but she does want to encourage us to think and do in ways that set us free from superfluity. She is a teacher above all and the book displays both insight and care for those who spend time in her company".

—Stephanie Donald, *Professor of Comparative Film, University of New South Wales, Australia*

"This is a timely and important book in a climate of increased recognition and valuing of creative practice research in the academy. Written by someone who not only *knows* this world, but also *does* this world, it is a rich, insightful and playful account of how mobile media are influencing the lives of creative practitioners—and how creative practices are emerging from mobile media users. A truly wonderful book for researchers and practitioners working across a range of disciplines, from anthropology to ethnography, and creative writing to screen production".

—Craig Batty, *Director, Higher Degrees by Research, School of Media and Communication, RMIT University, Australia*

"Marsha Berry's compelling new book explores the ubiquity of mobile media and its symbiotic relationship with human everyday practices and sociality. Berry argues that the networked and hybrid (online/offline) lives we now live provide new opportunities for extending both our creative practices and social circles, a digital futurism that is neither dysphoric nor utopian. This book is a powerful contribution to both creativity and ethnography scholarship, and simply a darned good read! Go buy it".

—Anne M. Harris, *ARC Future Fellow and Vice Chancellor's Senior Research Fellow, RMIT University, Australia*

Marsha Berry

Creating
with Mobile Media

palgrave
macmillan

Marsha Berry
School of Media and Communication
Royal Melbourne Institute of Technology
Melbourne
VIC, Australia

ISBN 978-3-319-65315-0   ISBN 978-3-319-65316-7   (eBook)
DOI 10.1007/978-3-319-65316-7

Library of Congress Control Number: 2017950697

Cover credit: © Morsa Images/DigitalVision/Getty Images

Printed on acid-free paper

This Palgrave Macmillan imprint is published by Springer Nature
The registered company is Springer International Publishing AG
The registered company address is: Gewerbestrasse 11, 6330 Cham, Switzerland

ACKNOWLEDGEMENTS

I would like to acknowledge the support of the former Design Research Institute, RMIT University as well as the School of Media and Communication and the Digital Ethnography Research Centre (DERC), RMIT University. I would also like to offer special thanks to Associate Professor Margaret Hamilton who worked together with me and provided valuable insights and knowledge from a computer science and information systems perspective for numerous mobile media projects.

I would like to thank Stefan Schutt, Peter Wilkin, Benjamin Solah, Paul Amoz Tan, Marion Piper, Jenny Weight, Lucinda Strahan, Maddy Morris, Dean Keep and Mel Jepson for their generous participation in Twitter poetry projects.

I would like to offer special thanks to Professor Jo Tacchi for suggesting me to write a book that would draw together a decade of my creative practice and ethnographic research into mobile media in the first place, and also to Professor Heather Horst, Distinguished Professor Larissa Hjorth and Associate Professor Craig Betty for their generous encouragement of this project.

Wholehearted thanks are also due to Glenn Ramirez and Shaun Vigil and the rest of the support team at Palgrave.

And last but not least, I would like to thank my children (Nicholas and Natalie) and good friends for listening to me and caring about my ideas.

Contents

LIST OF FIGURES

INTRODUCTION

.

Imagine there are no smartphones. Imagine your days without access to social media, family, friends and school or work as you move around about your daily business. How would you manage your life? Mobile media is everywhere—literally. An almost symbiotic bond has developed between our smartphones and ourselves. Think about all the things you do most days with your smartphone: talking to people, texting, commenting on social media posts, taking a selfie, posting photos of reflections, clouds and sunsets, sharing cat memes and videos, playing games, reading news articles, sending calendar alerts, responding to emails and so the list continues. No smartphones would mean we would be cast adrift from many those we are connected with through networked technology. Much of what we take for granted would require a special effort on our part. We'd lose many of the conveniences to which we have become accustomed. We'd also lose a source of constant entertainment as well as communication. But this presents a somewhat idealized picture of our relationship with mobile technology—symbiosis is a relationship of co-dependency that can have a dark side, after all.

The assemblages of mobile media are double-edged and not without popular controversy. Like many technological advances, they have a dark side as well providing us with new benefits and conveniences. Media commentators regularly decry the time we spend being connected, the anxiety we feel when we are disconnected, the ways in which life has sped up, the lack of work and life balance and the lack of time we feel we have to engage with the reasons we work—home and family. On the other hand, we are bombarded with advertising that exploits our fears of being unconnected and falling behind in a world giddy with technological consumerism. It is easy to persuade people that life was better in a "before" where there were no wireless mobile computing and demanding social software such as Facebook and Twitter: when everyday life and social interactions were not so mediated. Newspapers carry claims that the pressure to be constantly available and connected causes stress and actually diminishes the ability to concentrate with some media reports arguing that mobile media is rewiring our brain (Bea 2013, n.p.). We are perpetually connected to our phones. Media scholar, Sherry Turkle (2008) calls them virtual leashes that constrain and control us whereby "we are tethered to our 'always on/always on us' communication devices and the people and things we reach through them" (2008, 2)—yet there is evidence that they have also fuelled creativity by providing new outlets for creative impulses.

The central thematic focus of this book is on how mobile media has created new opportunities and contexts for creative practitioners. This notion provides a centre of gravity for this book where I draw together my own creative practice research with ethnographic projects I have undertaken. Over the past decade (2006–2017), I have explored how mobile technologies insinuate themselves into everyday social activities and rituals to become part of daily routines, and how artists and writers can use networked technologies to extend both their creative practices and audience reach. I am interested in how the lines between online and offline everyday life become blurred through both smartphones and mobile media.

My methodology dances in the space between creative practice research and digital ethnography. I have avoided a hard and fast distinction between online and offline following prominent digital ethnographers, Hjorth and Pink (2012). What lies between creative practices, and mobile and social media remains under-theorized and it is my hope that this volume goes part of the way to address this gap. The creative practices I explore include improvising, performing and playing with vernacular as well as with poetic forms of creative expression.

My approach to these creative activities and practices is inductive, intuitive and is influenced by non-representational theory (Ingold 2011) and ethnography as well as creative practice research paradigms and debates. Creative practice research has been concerned with theorizing the actions of making of creative artefacts as research processes in ways that are acceptable to the academy. Theoretical frames are often drawn from cognate discipline areas, for example, cinema studies often inform film-making and literary studies frequently underpin creative writing research. This can place the emphasis on the representational and signifying aspects of the artefact rather than on the activities that make up the practice itself. Such approaches present the danger of losing the richness and dynamism of the practice. Hence, there is considerable soul searching and debate about how to situate and write about creative practice as research amongst practitioner academics (see, e.g. Smith and Dean 2009, Batty and Berry 2016). This volume contributes a fresh approach to this debate by bringing together non-representational theory, ethnography and creative practice through the lens of everyday creative activities with mobile media.

The more than textual, more than representational, material and corporeal experiences are important to how we imagine and theorize creative practices with mobile media. I use non-representational theory as a theoretical perspective for the various actions such as writing poetry, taking photographs and shooting video that make up creative practices undertaken with smartphones and mobile media. Non-representational theory in essence is concerned with the background, the often invisible and slippery aspects of the enacted: the embodied, animate, and dynamic dimensions how we do things in our lifeworlds. Yet these very things are not often discussed or acknowledged in academic literature. Anthropologist, Tim Ingold in his preface to his volume of collected essays called *Being Alive* expresses this concern in the form of a question with a self-evident answer:

> Why do we acknowledge only our textual sources but not the ground we walk, the ever-changing skies, mountains and rivers, rocks and trees, the houses we inhabit and the tools we use, not to mention the innumerable companions, both non-human animals and fellow humans, with which and with whom we share our lives? They are constantly inspiring us, challenging us, telling us things. (Ingold 2011, xii)

This raises a further question that can be inferred from Ingold's question—How do we acknowledge our lifeworlds as sources of inspiration and knowledge when we write for the academy? I propose that a partial and tentative answer to this conundrum lies in a non-representational approach and can also lie in creative practice research where the focus is on the material aspects of the processes that make up a practice of creating and making rather than on symbolic and representational textual analysis of the creative outputs of that practice. In this book, I have unfolded the actions of creative practice through vignettes drawn from my ethnographic research to show how these manifest in our everyday lifeworlds.

Our social media and creative practices have become entangled with our daily routines, and this has changed the landscape of cultural production. Socializing, image making, creative writing, wayfaring and place making can occur simultaneously in online and offline worlds. Mobile media have enabled a blossoming of new creative vernaculars as well as an expansion of established fields of creative arts endeavours including photography, video art and poetry. The sheer volume of photos, video and poetry provides ample evidence of people's desire to document and share even their most ordinary moments of life in evocative ways. Phones have become bound closely to our aesthetic sensibilities and entangled with our muses. They shape how we find expression for our aesthetic impulses and provide us with tools and outlets for our creative practices.

Creative practices are made of actions and processes and have the intention of inventing and making creative works with and for mobile media. My own creative practice encompasses video, photography and poetry. The actions undertaken within my practice include creative writing, shooting still and moving images. The seven chapters present seven faces or aspects of creative practice that have manifested in both my ethnographic research and creative outputs in the past ten years. The essence that binds them together is the notion that mobile and smartphone assemblages have created new ecologies for creative practice. This notion began as a hunch. I planted some seeds and began to water them through my actions—playing with image making, playing and improvising in social media spaces, finding fellow travellers and collaborating with them and sharing my work. I became curious about how my fellow travellers felt and thought about what they were doing in these social and mobile media spaces. I saw an opportunity to investigate how social and mobile media provided new ecologies for creative practices. My curiosity led me to conduct digital ethnographic research.

This book is a fresh look over the various research projects I have executed and revisited to educe common patterns and trajectories over the past decade or so. Some of the material presented here has been used as the basis for journal articles, but I have combined and re-presented the material in a fresh way for this book. These previous articles are noted and acknowledged where necessary. The chapters in this book have been arranged into a sequence according to the kinds of actions that make up various creative practices. They also may be read out of order as each chapter is relatively self-contained and has its own list of references.

Chapter 1 began as an imagining and a way of explaining my bricolage methodology at seminars designed to explore emerging creative practice research paradigms. Creative practice research had already adopted methodologies drawn from other disciplines such as education and with its reliance on reflective practice and action research in its various iterations. My thinking was that if one can adapt Donald Schön's concept of a reflective practitioner for creative practice research, then surely ethnography could be invited to the discussion as well. What if creative practice and ethnography met? What would transpire between them? How can these research strategies inform and inspire each other? I thought about what lay between these two different research strategies with their origins in quite different disciplines and traditions.

I thought about the numerous conversations I have had with my doctoral students about the dilemmas of writing up creative practice research. Writing up ethnographic fieldwork is not straightforward either. In an earlier conceptualization of this chapter, which will appear edited volume later in 2017, was inflected by how ethnographic ways of working can be harnessed specifically by scholar film-makers who wish to have their creative practice considered as research. In this more comprehensive and conceptually complex version, I set the scene for how both digital ethnography and creative practice research with mobile media can benefit from each other's world views and methodologies. Here, I decided to animate my imaginings and conceived ethnography and creative practice research as two characters—Sarina as ethnography and Jed as creative practice—who meet and find that they have a mutual attraction and much in common to explore imaginary conversations and interactions between the two approaches to lived experience. In this chapter too, I lay down threads that continue throughout the book as I braid my own ethnographic and creative practice research into my discussions using semi-fictionalized and fictional vignettes based on the material I have collected in my fieldnotes and through interviews with my research participants.

Chapter 2 started life as a conference paper presented at ASPERA (Australian Screen Production, Education and Research Association) in 2014 where I introduced the idea that mobile media are creating conditions for the emergence of new visual creative vernaculars and that these extend to film-making. The conference theme was concerned with expanding practices. The paper was well received and generated some wonderful discussions. I was encouraged to submit an expanded version to a special issue of *Studies in Australasian Cinema*. I would like to thank the peer reviewers who asked challenging questions about my use of a non-representational theoretical approach that only made my paper stronger. Chapter 2 is the third evolution of this paper and includes substantial additional material drawn from my ethnographic research that did not appear in the journal article. In this chapter, I explore how creative practitioners who participate in social media groups play with photographs and video on an everyday basis.

Chapter 3 is completely new. I draw on my ethnographic research comprising observation and interviews to provide insights into how the self is presented and performed through social media. In it, I explore the phenomenon of selfies, self-presentation and immediacy in online contexts. I draw on my research to provide analyses of how online personas are enacted and how update statuses can reveal different stories about the dramas in people's lives. I engage with existing literature and media commentary to explore the selfie phenomenon from a relational perspective and ask: When we post selfies? What do we want to see? The practices examined in this chapter are concerned with photography.

Chapter 4 is a practical encounter between ethnography and creative practice research as it manifests in my work—both ethnographic and creative. An earlier and simpler version of this chapter was presented at the ASPERA (Australian Screen Production, Education and Research Association) in 2016 where I discussed what "being there" looks like in a lifeworld where the mobile media and digital co-presence is imbricated within everyday activities and rituals. I received some really valuable feedback about what people wanted to know about using non-representational approaches for their own research as well as great stories from audience members about how mobile media and digital co-presence figure in their lives on a daily basis. In this chapter, I have extended my thinking towards my own creative practice as well as what my ethnographic research was telling me. This chapter presents my thinking about my creative experiments and art making with mobile media. I have written about using ethnographic techniques and in the spirit of non-representational approaches to writing for the academy. This chapter's centre of gravity is mobile video and photography and the actions needed to create these with and for mobile media.

Chapter 5 has its origins in an article I wrote for Text Journal, called "Poetic Tweets", which was published in 2011. In that article, I analysed the implications of Twitter for creative writers and examined how they were using Twitter to engage with each other as well as with their own practice. For example, through Twitter, the reach of National Poetry Writing Month held in April in the USA has transcended national boundaries and become an international event (NaPoMonWri 2011). Many took up the challenge to write a poem each day through April. Blogs arose to provide stimulus words and forums to support poets who chose to participate. The connection to Twitter provided individual blogs with immediacy so that people can publicize their creative practice to others who care and are interested. Writers receive the benefits of belonging to an expanded network of peers as well as the opportunity to perform in virtual "salons" to an appreciative audience. When I wrote "Poetic Tweets", these practices were relatively new. By the middle of the second decade of the twenty-first century, these practices had spread and evolved further and continue to do so. In this chapter, I revisit the material I collected in 2010 and wrote about in "Poetic Tweets" and add new material from a participatory mobile art project I directed in 2013. The actions that make up creative practices in this chapter are concerned with performing, improvising and writing poetry in mobile media ecologies.

Chapter 6 is the offspring of my ongoing research project with pinning community-generated poetry to Google maps. The project itself is housed on a blog—*Poetry4U* that my colleague Omega Goodwin and I maintain. I have previously written about this project for the journal *New Media and Society*, and I would like to thank the editors and peer reviewers for their erudite feedback that helped me synthesize the plethora of literature theorizing place, space and landscape. *Poetry4U* has taken me on some wonderful journeys including to the Pilbara in north-west Australia. I wrote this fieldtrip up for a special issue of *Coolabah* focusing on the corporeal and embodied aspects of being in a landscape. Here, I present the current state of play for *Poetry4U* as a participatory locative media art project through three sub-projects. This chapter spirals around the actions, processes and strategies used to gather poetry for mobile media that engages with a sense of place from creative writers.

Chapter 7 first appeared as an essay in a volume of collected essays, *The Routledge Companion to Digital Ethnography*. My focus in that essay was on digital ethnography and the dynamics of mobile media ecologies. Here, the mainstay is creative practices, namely the actions of making films and video with mobile phones and their later iterations in the form of smartphones. I ask questions about how we can better understand the dynamic processes and actions associated with creating films and videos with smartphone assemblages which themselves are constantly evolving because of technological advances. I include reflective and autoethnographic accounts of the actions that were integral to the processes within my own video-making creative practice over the past decade. This chapter includes careful deliberations on a decade video and film-making practices with mobile media that situates these practices within a genealogy of cultural practices.

Chapter 8 revisits the overarching themes of the book: the emergence and evolution of new visual vernaculars; how we use selfies as a remediation of self-portraits and postcards; how we imagine and reimagine places and events through the use of smartphone camera to evoke a sense of "being there"; how improvising and collaborating with other writers in social media spaces takes on a performance as well as a social dimension; how location-based technologies offer the ability to create new cultural spaces; and mobile film-making, video and photography from a perspective informed by creative arts practice research as well as ethnography. It draws together how ethnographic approaches with smartphones and other mobile devices can extend creative practice research into new directions.

Overall, this book investigates the convergence between locative, social and mobile media by layering creative practice research with non-representational theory and digital ethnography to show how people who identify as creative practitioners are using mobile media for their creative practice—creative writing, photography, video and film-making.

REFERENCES

Batty, Craig and Berry, Marsha. 2016. Constellations and connections: the playful space of the creative practice research degree. *Journal of Media Practice* 16 (3) 181–194, http://dx.doi.org/10.1080/14682753.2015.1116753.

Bea, Francis. 2013. The Internet is Rewiring Your Brain and You Don't Even Know It. https://www.digitaltrends.com/social-media/the-internet-is-rewiring-our-brains-we-just-dont-realize-it/. Accessed May 5 2017.

Ingold, Tim. 2011. *Being Alive: Essays on Movement, Knowledge and Description.* New York: Routledge.

Pink, S and Hjorth, L. 2012. 'Emplaced cartographies: reconceptualising camera phone practices in an age of locative media'. *MIA (Media International Australia) Incorporating Culture and Policy: quarterly journal of media research and resources,* 145: 145–155.

Smith, Hazel and Dean, Roger. 2009. *Practice-led research, research-led practice in the creative arts.* Edinburgh: Edinburgh University Press.

Turkle, Sherry. 2008. http://web.mit.edu/sturkle/www/Always-on%20Always-on-you_The%20Tethered%20Self_ST.pdf. http://dx.doi.org/10.7551/mitpress/9780262113120.003.0010.

CHAPTER 1

Creative Practice Meets Ethnography

AN ENCOUNTER

Please allow me to set the scene for how both digital ethnography and creative practice research with mobile media can benefit from each other's world views and methodologies. Let me begin with the bones of a fictional vignette.

Jed walks into a café near the university where he has recently commenced a PhD by creative practice. It's decorated in a ubiquitous inner urban hipster style with upscaled junk. He spies Sarina sitting at the bench table by the window; she's looking intently at the street and sketching something in a notebook. Jed narrows his eyes and tries to see the page in her notebook. It's too far away. He orders a long black coffee at the counter, pours a glass of water and joins Sarina by the window. She's scribbling on a page in her notebook facing the sketch of the streetscape that had initially grabbed his interest. He can't read the handwriting quickly enough—but the layout suggests a poem of some sort. He catches her eyes, "Hello", he says and smiles. Greetings exchanged, she closes her notebook. Small talk commences. She thinks he's cute. Turns out they're both doing doctoral studies at the same university. Sarina's PhD is all about the selfie phenomenon and she is using ethnography. She's enrolled in the Anthropology department. She has been filming pedestrians with her smartphone in public places and busy sidewalks as an initial step in her research. Jed wants to probe deeper and see what she is doing in her journal. Time for the reciprocal disclose. He's a mobile media artist and is looking for conceptual and

© The Author(s) 2017
M. Berry, *Creating with Mobile Media*,
DOI 10.1007/978-3-319-65316-7_1

1

methodological frames. He's enrolled in the Digital Media department. He tells her about his journal. Takes it out of his bag and shows her a couple of pages. Reflections, storyboards, mind maps. Messy. He's exploring locative mobile media and is thinking about making a walking trail around some of the urban laneways with place-based historical and contemporary narratives. Each place on the walking trail will have layered stories, fictional and nonfictional. Their eyes lock. She nods, blushing. She opens her notebook and shows him the sketch and the poem she's just written. She tells him about the need to be there in the field site – this is part of her site, this street, this café, and this suburb. She's looking for connections – between places and selfies, between walking and selfies, between what people say and what they do with selfies, between selfies and street art... the list continues. He nods. Clearly, there is chemistry between them.

In this chapter, I am concerned to explore the connections between creative practice research and ethnography. I ask: How can they inform and benefit each other? In terms of the academy, ethnography has a long pedigree whereas creative practice research is still establishing itself in its own right. Creative practice research is a bricolage where methods such as participatory action research are appropriated and adapted to fit into the world views and epistemologies that underpin creative practice research. On the other hand, ethnography claims to be about writing, about translating the experience of fieldwork into writing. Notions of immersion in a particular site and participant observation underpin fieldwork itself where the researcher is aware of his/her responses as well as potential impacts on the social interactions and activities within site itself in a reflexive way. The experience is analysed, interpreted and written up following academic genre conventions and theorized using conceptual frames and structures drawn from anthropology and sociology. Ethnography draws its ontology and epistemology from anthropology and uses research strategies and techniques such as immersion, insider perspectives, participant observation and interviews.

In their text, *Writing Ethnographic Fieldnotes* (1995), Emerson, Fretz and Shaw observe that the work of ethnography involves two activities. First of all, the ethnographer enters a social setting and becomes familiar with various social activities and daily routines. Secondly, the ethnographer documents their experiences into an accumulative record, including notes, sketches, photographs, videos, screenshots and interview recordings. Immersion has long been a key dimension of ethnographic work. Goffman famously stated that research involves "subjecting yourself, your own body and your own personality, and your own social situation, to the set of contingencies that play upon a set of individuals, so that you can physically and ecologically penetrate their circle of response to their social situation, or their work situation, or their ethnic situation" (Goffman 1989, 125). The work of the ethnographer may be likened to a hero's journey where the ethnographer enters the unknown and returns with the elixir.

James Clifford and George Marcus famously problematized the act of writing up fieldwork (1986) whereby the writing itself was no longer regarded as neutral or objective. They challenged ethnography as a representational enterprise in the 1970s and 1980s with postmodern approaches that exposed the politics of representation and established that point of view and the style of writing have a profound influence on the way research is read and received. They argued that ethnographers should identify their own subject position explicitly and reflexively within the account of the research instead of relying on a third person perspective to create an illusion of objectivity and impartiality. They advocated a polyvocal approach to write that would allow for the incorporation of the individual voices and subject positions of participants rather than subsume them under the unitary authoritative voice of the researcher. The postmodern crisis in representation has led to some anthropologists such as Kent Maynard (2002) to embrace poetry as a way to write ethnography. The Society for Humanistic Anthropology includes installations at its annual conference and holds annual ethnographic poetry and fiction competitions.

The debates around representation and structure in how to write about ethnographic fieldwork contribute discussions about how to write about creative practice research because each of these discipline areas is seeking ways to write a about lived experience with all its messiness and entanglements in non-reductive ways that will satisfy the rigors and requirements of the academy. There is a valuable alignment here between creative practice and ethnography because the mainstays of creative writing include forms and genres such as memoir, life writing, lyrical non-fiction essays, short stories and literary novels to mention but a few and screen production has a long history and tradition of documentary filmmaking. These traditions and forms can be appropriated for ethnography. Clearly, there are useful connections, associations, conjunctions and juxtapositions to be made between how writing and visual storytelling are thought about across these two discipline areas.

Ethnographers move between immersion and distance while being in the field. Effective field notes will often contain evocative and emotional accounts of events and happening and speculations as to possible meanings as well as detailed observations. Ethnographic writing is a mix of narrative and analysis, and the stance is both participatory and detached at the same time. And it is expected that the "being there" (Geertz 1988) will come across in the writing and through the writing. Ethnographic writing inscribes and enacts the reflexivity of the ethnographer as the writing moves between states of immersion and detachment, between passages containing thick descriptions and dramatized recounts of events, and the analytical passages theorizing social structures and relations through careful triangulation of evidence. Lived experience with its contextual inflections and nuances and the conundrum of attending to the corporeal and multi-sensory aspects of lived experience are critical concerns of contemporary ethnography.

The sensory turn in ethnography according to Pink (2009) seeks to "attend to and interpret the experiential, individual, idiosyncratic and contextual nature of research participants' sensory practices and also seeks to comprehend the culturally specific categories, conventions, moralities and knowledge that informs how people understand their experiences" (Pink 2009, 15). The sensory turn emphasizes the embodied nature of an ethnographer's experiences and the importance of acknowledging these with a conscious and self-reflexive awareness. In her own ethnographic practice, Pink (2009) used video to document how her participants go about daily routines such as laundry and housework in their homes. She uses this material to step inside such routines in a reflexive way to rethink our relations and with inanimate material objects to include sensory aspects.

Reflexivity or "thinking from within experiences" (Bolton 2010, 14) is fundamental to ethnography. It's more than reflecting back on an experience and questioning what happened, why, what did I think/feel and how would I do it differently, etc., which we tend to associate with action research. The notion of reflection as a process integral to develop practices is usually attributed to John Dewey who argued in *How We Think* that experience alone does not teach us things, it is only by reflecting on an experience and information with questions about the relationships and meaning that we begin to learn from that experience. Reflection is really about what happened or was done in the past. Reflexivity is about being consciously aware of experience in the present, as it is happening.

The Oxford dictionary defines reflexivity as "turned or reflected back upon the mind itself". It is an approach to experience and epistemology where one seeks to be aware of one's own habitual patterns of thought, world view and interpretations and to step back from these so that one is able to observe one's internal dialogue, where it's come from and what is framing it and how it is affecting an experience in a moment of time. It is a state of being that is incredibly useful for scholarly creative practice as well as essential for ethnography. Fook describes reflexivity thus:

> Reflexivity is a stance of being able to locate oneself in the picture, to appreciate how one's own self influences [actions]. Reflexivity is potentially more complex than being reflective, in that the potential for understanding ways in which one's own presence and perspective influence the knowledge and actions which are created is potentially more problematic than simple searching out for an implicit theory. (Fooke 2001, 43)

The notion of reflexivity includes a conscious awareness of one's lifeworld or what Husserl calls the *Lebenswelt*, which has been aptly defined by the Merriam-Webster dictionary as "the sum total of physical surroundings and everyday experiences that make an individual's world" (https://www.merriam-webster.com/dictionary/lifeworld). To dig deeper into the notions of reflexivity and lifeworld, I turn to Husserl's essay (1954/1970), *The Crisis of European Sciences and Transcendental Phenomenology* where he explains:

> The world exists as a temporal, a spatiotemporal, world in which each thing has its bodily extension and duration and, again in respect to these, its position in universal time and in space. It is as such that we are ever conscious of the world in waking consciousness, as such that it is valid as universal horizon. Perception is related only to the present. But this present is always meant as having an endless past behind it and an open future before it. (63)

And later in the same work he explicates:

> For the lifeworld — the "world for us all" — is identical with the world that can be commonly talked about. Every new apperception leads essentially, through apperceptive transference, to a new typification of the surrounding world and in social intercourse to a naming which immediately flows into the common language. Thus the world is always such that it can be empirically, generally (intersubjectively) explicated and, at the same time, linguistically explicated. (99)

Lifeworlds of individuals are intersubjective, dynamic, relational and much more than representation. They are a key conceptual underpinning for non-representational theory and provide way to move beyond the politics of representation to embrace vitality, fluidity and as hybrid relations we have with the physical world and mobile media. The non-representational turn in ethnography can offer researchers new ways of framing creative practice research. In the Foreword to *Non-Representational Methodologies: Re-Envisioning Research* edited by Phillip Vannini (2015b), the world-renowned anthropologist, Tim Ingold urges academic researchers to embrace complex sensory lifeworlds and to take risks with conceptualizing and communicating their discoveries (including with academic writing genres) through non-representational theory:

> One night, a few years ago, I woke from a dream with the following lines in my head:
> Often in the midst of my endeavors
> Something ups and says
> "Enough of words,
> Let's meet the world".

> I do not know who put these lines there. Certainly, I did not invent them. But immediately upon waking, and before they had time to evaporate, I rose from my bed to write them down. They remain, pinned to a notice board in my office, and every so often I take a look at them, to remind myself of the message they contain. They could perhaps be taken as a manifesto for a non-representational way of working. (Ingold 2015, vii)

These words from Ingold also celebrate the importance and validity of working in intuitive and inductive ways, both of which are critical to creative practice research as well as to ethnography.

A SECOND RENDEZVOUS: NON-REPRESENTATIONAL APPROACHES

In a recent book chapter about ethnography and screen production (Berry 2017), I argued that non-representational approaches provide researchers in screen production with ways to move beyond the familiar paths of representation where the practice of film-making is subsumed by cinema studies of digital humanities. Here, I extend this to refer to the gamut of creative practices associated with smartphone assemblages including mobile and social media, film-making, photography and writing. Creative practice research is not just about form and artefacts and symbolic meanings, it's also about constellations of processes, connections and relationships. Nigel Thrift developed non-representational theory in human geography in the early 1990s (Thrift 2008; Vannini 2015a, b). It is an alternative to representational theories that privilege forms and objects. It is an "umbrella term for diverse work that seeks to better cope with our self-evidently more-than-human, more-than-textual, multisensual world" (Lorimer 2005, 83).

Let's now return to Sarina and Jed to see how their relationship is progressing:

> Sarina takes a sip of her craft beer. Jed is telling her about how he has been choosing sites for his locative media narrative. It's all about wayfaring, wayfinding and using smartphones as a portal into an alternate city. Players will be like flâneurs and will encounter geotagged locations where stories from the past and present can be accessed and where players can add their own impressions. He wants to experiment with smartphones as a way of intervening in the urban landscape to see what happens, and what kinds of stories will emerge from the participants. He tells Sarina about a locative media project he saw on Valentine's Day that was run as a marketing ploy by a major telco where people were using their smartphones to post messages to a large billboard in a public place. People were gathered around waiting for their own messages and didn't seem to pay any attention to other participants. He wanted there to be more of a connection between people in his game – a bit like the groups that gathered together to play Pokémon Go when it first came out. Sarina thought a minute. "Hmm, that's more than representation – that's about what people do with locative games and installations with smartphone assemblages and how they feel about it".

Sarina understands that Jed's practice includes observations of how people use smartphone assemblages in public places. Although observing how people engage with media art projects isn't directly his part of art practice, it is integral to his project. An installation with an urban screen adds an affective atmosphere to public place that lifts it out of the ordinary. Sarina saw that non-representational theory could uncover some practical methods for Jed to work with his observations in the vignette above. Events, happenings and unfoldings are focal points for non-representational ethnography because events bring forth dramas, repetitions, uncertainties and disruptions. They offer opportunities to find new ways of thinking about subjectivities, differences and repetitions that take on the form of rituals. It is important to attend to relations because life arises from the entanglements of actors. As Anderson and Harrison observe, non-representational ethnography examines spaces "where many things gather, not just deliberative humans, but a diverse range of actors and forces, some of which we know about, some not, and some of which may be just on the edge of awareness" (Anderson and Harrison 2010, 10). Non-representational theory is concerned with "corporeal rituals and entanglements in embodied action rather than talk or cognitive attitudes" (Vannini 2015b, 4). It provides a course of action to explore the dynamic, the fluid, the corporeal, the material, the ineffable and the ambiguous.

Ethnographers using non-representational approaches explore associations, mutual formations, ecologies, constellations and co-fabrications. This approach to methodology can be of benefit to Jed because he also seeks to examine thought in action through doings, practices and performances. Performance and performativity are integral parts of the locative media events described by Jed in his second rendezvous with Sarina. To illustrate further how concepts and outlooks drawn from anthropology and ethnography can benefit creative practice research in mobile media, I turn to Schechner (2002), an influential theatre anthropologist and performance studies scholar, who sees performance as a possibility that waits for the unfolding of practice. Movement and affective resonances are critical to such unfoldings. All of that which sits in the background unspoken, which includes tacit knowledge of social norms and how ritual interactions should go, physical conditions such as the weather and affective atmospheres are important because these are "open to intervention, manipulation, and innovation" as well as to "colonisation, domination, control, cultivation and intervention" (Anderson and Harrison 2010, pp. 10–11). These entanglements reverberate across the lifeworld to inform and shape each other and unfold into more intricate patterns to reveal new research directions in ethnography. Rather than trying to unpick them into component parts, non-representational approaches to methodology seek to understand how these entangled forces work by understanding them as complex meshworks (Ingold 2011). This perspective is also beneficial for better understanding the interlaced webs of actions and processes that may coalesce within a creative practice. Yet it is advantageous to make a distinction between an over-riding methodology and the methods.

Vannini (2015b) notes that methods and methodology are not the same. He is not the first to do this but he provides some very clear and useful distinctions that are beneficial for thinking about creative practice as research as well. He suggests a threefold approach whereby methods, research strategies and methodology are each thought of separately, and I suggest that his threefold approach is equally useful for those undertaking creative practice as research in all its different guises. Research methods are actually "procedures for the collection of empirical material" (Vannini 2015b, 10) or data. Methods such as interviews and visual dairies are the tools we have at our disposal to collect the material or data we need to address our research questions or to interrogate our research propositions. Research strategies are procedures for the treatment of data or a way to analyse and interpret the data we have collected. These may include reflections, narrative analysis, content analysis, participatory action research or "very much anything else that fits the researcher's preference" (Vannini 2015b, 11). Methodologies are the "epistemological foundations" (Vannini 2015b, 11) for the research.

The goal of non-representational theory as an epistemological underpinning for research then is to embrace practice, embodiment, materiality, and process or the more than representational aspects and phenomena. It seeks out entanglements and relations we have with objects rather than their structures and symbolic meanings (Vannini 2015a, 320). Non-representational approaches accommodate conceptual portals and cartographies for Jed to explore the relations between smartphones, people, places and contemporary and historical narratives through a locative media project.

Non-representational theory has been a major influence in contemporary anthropology (see for example the work of Tim Ingold, Sarah Pink and Phillip Vannini). Vannini notes that "non-representational theory seeks to cultivate an affinity for the analysis of events, practices, assemblages, affective atmospheres and the backgrounds of everyday life against which relations unfold in their myriad potentials (Vannini 2015a, 318). This resonates and overlaps with ethnography which "involves the ethnographer participating, overtly or covertly, in people's daily lives for an extended period of time, watching what happens, listening to what is said, asking questions—in fact, collecting whatever data are available to throw light on the issues that are the focus of the research" (Hammersly and Atkinson as cited in Pink 2001, 18). Also, this way of working could be really useful for our fictional locative media artist Jed. Furthermore, non-representational ethnographers, according to

Vannini, "consider their work to be impressionistic and inevitably creative" (2015a, 318). And this has strong resonances with practice-led and practice-based research where the poetic and the aesthetic matter. Clearly, there is reciprocity here—creative practice researchers have much to offer ethnographers.

To pursue what non-representational research might look like in creative practice research contexts further, I present Vannini's formulation of five qualities he calls an "ethos of *animation*" (2015a, 319) which characterize non-representational methodologies. He commences with the quality of vitality (2015a) where he observes, "A vitalist ethnography, in short, is an ethnography pulled and pushed by a sense of wonder and awe with a world that is forever escaping, and yet seductively demanding, our comprehension" (2015a, 320). Intuition and inductive ways of working that embrace the evocative and the affective reflect an ethos of animation. This is incredibly useful for creative practice research and opens a myriad of possibilities for methodological synergies and complementary relations between ethnography and creative research.

The second quality is performativity whereby there is a focus on action, which "emphasizes the importance of ritualized performances, habitual and non-habitual behaviors, play and the various scripted and unscripted, uncertain and unsuccessful doings of which everyday life is made" (2015a, 320). This has resonance with the oeuvres of Erving Goffman and Richard Schechner where both performance and performative aspects of interactions are emphasized. Performativity often underpins creative practice research as well and Haseman (2006) has written an eloquent manifesto advocating the importance of performativity as a quality of research involving the creative arts, and I address this in later in this chapter.

The third quality is corporeality, which underscores the notion of "the researcher's body as the key instrument for knowing, sensing, feeling and relating to others and self" (2015a, 321) so that "affect is a medium through which ethnographic research unfolds" (2015a, 321). Arguably, corporeality is essential for creative practice research as well where the practice itself is central to the research enquiry. So again there are clear synergies between a non-representational approach drawn from ethnography, especially in its sensory manifestations, (see Pink 2009) and creative practice research.

The fourth quality identified is sensuality and this serves to "underline the not-necessarily reflexive sensory dimensions of experience by paying attention to the perceptual dimensions" (2015a, 322). Sensuality and perceptual dimensions are key concerns for creative practice research as well. The fifth and final quality is mobility, which seeks to account for the kinetic dimensions of fieldwork where "ethnographic journeys are not planned transitions from the office to the field site but wanderings through which movement speaks" (2015a, 323).

The "ethos of *animation*" as formulated by Vannini (2015a, 319) is well suited to inform creative practice research methodologies because it embraces diverse ways of knowing as well providing a way to account for the multi-sensory and affective dimensions. Yet this relation is not necessarily unilateral, rather if we conceive of it as reciprocal then the field of ethnography may be further expanded through creative practice methodologies, strategies and methods. I return to the story of Sarina and Jed's growing friendship to illustrate this point.

Sarina stands on a street corner filming people walking in a busy street in a city. The people are zombies, she thinks, they're are more concerned with the contents of their smartphone screens than what is happening in front of their noses. She's collecting material for her doctorate. It occurs to her that she could make a video art work as well. That evening she wrote up her impressions in the form of rough field notes. She wondered at the way homeless people were invisible to most passers-by. She reviewed the video she had shot. Some of it was hilarious and some very sad. Maybe Jed can have a look and see how best to thread it into a narrative form. It would be fun to present it at the conference and maybe it could be part of the exhibition the creative arts students were planning.

She started thinking about a paper she would present next month at a conference on mobile media. She wanted it to be a lively animated account that would evoke as many of the sensory aspects of her observations as possible yet would still be deemed as scholarly by the academy. She wondered whether she should write a plain or straightforward descriptive and analytic account or whether she could take a risk and compose an impressionistic sensual account by creating a fictional digital wayfarer (Hjorth and Pink 2014) who meanders along the busy streets of the city avoiding any direct eye contact with fellow pedestrians. Alternatively maybe she could start with free writing and see what comes up. She opened a fresh document and stared at the white screen.

Situating Creative Practice Strategies and Methods

Before I explore how strategies drawn from creative practice can assist Sarina, I would like to outline some key debates in creative practice research circles that are of relevance to synergistic and reciprocal relations between creative practice research and ethnography. In 2009, Grierson and Brearley wrote that "there is a growing need for the articulation of research methodologies appropriate to creative arts practice" (p. 4) in the introduction for a collection of essays they edited drawn from a creative practice research strategies course. The selection of methodology is never neutral. Since 2009, there has been a consistent growth in research aimed at articulating the shapes and forms of creative practice research (e.g. see Batty and Berry 2015; Berry and Batty 2015).

The relationship between theory and practice is often fraught for practice research discourses yet this relationship is critical to understanding how the actions that make up a creative practice may become research. Brabazon and Dagli see the relationship as a one that can "create a dialogue between theory and practice, to raise questions that cannot be raised within practice, to probe the applications within the theory and/or to follow the process of thought in order to identify the intellectual pathway in/to the creation of visual [or other] propositions" (Brabazon and Dagli 2010, 36–37). Theory provides us with systematic ways to think about the activities that make up practices and provide reasons, interpretations and explanations of these.

In previous work, (Berry 2017) I have suggested that creative practice research in the screen production areas tends to be preoccupied with representation. Film-making, photography and screenwriting suggest representational research strategies with a strong focus on film and video artifacts as texts. However, over the last decade, there has been an increased push in creative practice research towards performative research that emphasizes the doing and the making rather than on the interpretation of symbolic meanings and authorial intentions. This brings me back to a point I made at the outset of this chapter where I conceived of creative practice research as a bricolage that poaches strategies and methods from other fields and disciplines.

Donald Schön is a key thinker who has framed much of how we envisage creative practice as research. Schön framed his theory within the discipline of urban planning at Massachusetts Institute of Technology where he was a professor, and his work builds on that of the philosopher and educator John Dewey. He theorized that tacit knowledge arises from "the situations of practice—the complexity, uncertainty, instability, uniqueness and value conflicts which are increasingly perceived as central to the world of professional practice" (Schön 1983, 14) and drew attention to the fact that situation of practice and tacit knowledge is often difficult to articulate. Schön also developed a methodological approach based on the notion of a reflective practitioner who reflects back on experiences and situations of practice in order to engage in a process of continual learning. This methodological approach has become popular in design disciplines (Smith and Dean 2009).

Haseman (2006), who works in theatre and performance disciplines, draws on Schön's work to analyse and theorize practice-based research strategies and identifies the reflective practitioner, participant research, participatory research, collaborative enquiry and action research as techniques and methods that may be harnessed for creative practice research. According to Haseman, practice-based research is "concerned with the improvement of practice, and new epistemologies of practice distilled from the insider's understandings of action in context" (Haseman 2006, 100). In 2006, Haseman wrote *A Manifesto for Performative Research* in which he identifies three characteristics of creative practice research. First, he claims that "practice-led researchers construct experiential starting points from which practice follows" (101) rather than being problem or research question driven. Second, he asserts that practice-led researchers are not interested in writing about the research in academically acceptable forms drawn from more established disciplines, rather they seek to produce outputs and knowledge in forms that reflect their practice. Haseman proposes a third methodological distinction where findings may be presented in a rich presentational form such as an exhibition, a film or a performance rather than as a traditional doctoral thesis or dissertation. He argues that Austin's speech act theory (Austin 1962); especially the performative function of utterances where the utterance enacts what it names may be a way to understand how an exhibition, a film or a performance is the outcome of research. His interpretation sidesteps the complexities and the many nuances of Austin's speech act theory and to focus simply on the connection between a word and an action by claiming that for Austin "performative speech acts are utterances that accomplish, by their very enunciation, an action that generates effects" (Haseman 2006, 6).

He builds on this to argue that therefore research findings may be presented as performatives that perform an action, and this will place practice as the main research activity rather than the writing up of the research into academically acceptable genres. Clearly, Haseman is frustrated with the limitations of modifying existing research methods to accommodate creative practice as research as well as the need to write up one's research into a form recognized by the academy. He is also frustrated with the need to shoehorn creative practice research into forms that are driven by older research paradigms such as textual and discourse analysis as well as the familiar scientific methods with their emphasis on testing hypotheses. His argument and discussion hang on the shortfalls of specific methods rather than on the epistemological and ontological problems faced by researchers for whom research is a material, embodied and empirical process.

Much of the discussion around what creative practice research embodies and whether or not a film or screenplay may be seen as research outputs hinges around the relationship between theory and practice. There are reciprocal and iterative connections between theory and practice that have been classified into two categories by Smith and Dean (2009)—practice-led research and research-led practice. Practice-led research is understood as both "the work of art as a form of research and to the creation of the work as generating research insights which might then be documented, theorised and generalised" (Smith and Dean 2009, 7). Research-led practice complements practice-led research and implies that other forms of disciplinary research may be useful in informing creative practice research. It is creative practice that while initiated from basic research is not necessarily concerned with creative practice and can take "different forms in different fields" (Smith and Dean 2009, 8). For example, in screen writing research-led practice may be conceptually driven and informed by cultural studies and genre theory; in documentary filmmaking about a minority community, research-led practice may inform by the politics of representation and identity. In response to the distinction between research-led practice and practice-led research made by Smith and Dean (2009), Batty et al. (2015, 3) identify that

> … the conditions of creative practice research seem to fall into two categories. Firstly, there are those who use the research environment to better understand their practice, explicating what they know tacitly about the work they create. Secondly, there are those who use research to generate new ideas and concepts that either changes the way they practice (process), or that changes the fabric of their practice (content). (Batty et al. 2015, 3)

Research intentions, propositions and questions then are important considerations if we wish to gain insights into the forms creative practice research may assume. It is also quite possible that a research project may include an iterative cycle that dances between both research-led practice and practice-led research and between content and process and sheds light on a practitioner's tacit knowledge about his or her practice.

Creative practice research which is "supported by critical reflection and reflexive action can be seen to invert the research process because it encourages working from the unknown to the known" (Sullivan 2009, 48–49). Through creative practice research, we can question existing knowledge in order reveal "critical insights that can change what we do know" (Sullivan 2009, 48). We can also interrogate and theorize emergent cultural and artistic production manifesting in mobile and screen media contexts. Kerrigan et al. (2015) combined to write a landmark article in which they unearthed what creative practice research might look like where screen production is the research methodology itself. They concluded that research "in screen production is by no means a homogenous activity but usually involves the production of a film (or other screen work), an iterative process of practice and reflection by a researcher who is also the screen practitioner, and a theoretical perspective that informs the overall research" (Kerrigan et al 2015, 109). They also identified a clear need for more work in this area because of the diversity of both methods (techniques and strategies) and methodologies (ways of knowing).

Commonly held assumptions may be challenged when they no longer serve us through creative practice research. Our practice may be extended to new fields and forms, and we can document what the proliferation of technological advances offer us as film-makers, photographers, artists and creative writers because "the space of creative practice research encourages a critical engagement with doing, making, re-doing and remaking. It creates a place in which practice can be incubated alongside ideas, calling into question the past, present and future of that practice" (Batty and Berry 2015, 185). The space of ethnography creates a place where the social practices and material cultures of mobile media, creative writing and screen production can be better understood alongside through the lens of theories of social structures, customs, norms and rituals.

There is much written about creative practice as its own methodology. Nonetheless, the doing of creative practice can be written about using ethnographic ways of knowing as well as ethnographic processes and practices. Ethnographic ways of knowing are underpinned by an empirical approach that seeks accuracy and is grounded in material culture and social practices. The semantic field encompassing ethnography includes routines, social rituals, social interaction, exchange, material cultures, kinship, world views, belief systems and forms and drivers of social control and norm setting, initiation and rites of passage, systems of symbolic meaning—the list could go on forever to cover all aspects of human life. Creative practice similarly shares a broad focus. Such a broad focus is not conducive to disciplined study so a key challenge is to delineate the semantic field in a nuanced and specific way so that meaningful contributions to knowledge may be made. Ethnography values the nuanced narrative analytical accounts of lifeworlds. Details matter. Specifics matter. Contexts matter. Relationships matter, and what lies in between these matters. This epistemological and ontological orientation is conducive to creative practice research methodologies with their emphasis on inductive rather than top-down approaches.

Furthermore, when we look at what lies betwixt creative writing and ethnography we can see that creative writing techniques and pedagogies inform ethnography. Writing, and more specifically good writing, always has been at the heart of ethnography. Writing about known, familiar and taken for granted things can be made unfamiliar through what Geertz terms thick descriptions. A thick description is actually a type of writing that shows rather than tells or proclaims, and where the analysis emerges through the detailed descriptions rather than superimposed as a proclamation. Such writing when well executed is entertaining and gently leads the reader to the conclusions and insights rather than telling, qualifying, amending, modifying, defining, redefining and telling again as is common in academic genres. Geertz's famous work on Balinese cockfighting makes excellent use time honoured techniques such dramatic turning points as a narrative device as well as finely crafted thick descriptions to construct a compellingly detailed and accurate account of his field work in a Balinese village.

The anthropologist Kirin Narayan also draws on creative writing techniques and pedagogies. She lauds the power of free writing for writing ethnography. She is also an advocate of writing that is nuanced and argues for showing rather than telling when writing about one's research. She builds her case for the usefulness of this technique for ethnography by referring to Anton Chekhov's cannon of short stories, especially

Sakhtin Island, which is an account of a Tsarist penal colony. The work was never defined as ethnography yet it possessed many of qualities valued in strong ethnographic writing where passages of empirical precision sit alongside evocative description in an easy relationship. Vannini (2015a, b) also picks up on this theme when he discusses ways in which research can be written up that will appeal to popular audiences and raises the profile of academic research within mainstream media.

CONCERNING FICTION AND EXPERIENCE

Ethnography offers creative practitioners a way of moving beyond the familiar ground of representation, symbolic meaning and textuality as well as analytical text-based methods and strategies associated with humanities disciplines. Writing is central to ethnography, and ethnographic writing does have clear intersections with life writing, biography, autobiography, travel writing, lyric essays and ficto-critical genres. Ethnographic writing is concerned with both theory and practice and may be described as a mix of narrative and analysis to produce thick descriptions that capture the nuances of an event with the aim of conveying the experiential dimensions of what it is like to "be there" in a research site to a reader. Ethnography seeks to place specific aspects of life against "a cultural whole" (Baszanger and Dodier 2004: 13). It seeks to pay attention to the evocative, affective and sensory aspects of everyday life, particularly when a non-representational (Vaninni 2015a, b) or sensory (Pink 2009) approach is taken.

In previous work, (Berry 2016, 2017) I have referred to a model of writing ethnography developed by Humphreys and Watson (2009) that I have found useful for my own practice as well as for my research students. Their fourfold model presupposes that ethnography is an act of writing at a fundamental level and thereby attends to the problematic how to write up research. They analyse ethnographic writing into four forms: "the *plain*, the *enhanced*, the *semi-fictionalized* and the *fictionalized*" (2009: 41). I see this model as a bridge between ethnography and creative practice research and shows what happens when ethnography draws on creative writing techniques to investigate and communicate discoveries and insights made through fieldwork.

Plain ethnography, the first form in the model proposed by Humphreys and Watson (2009), is a traditional account of events informed by theory and contributions to existing theory are argued through the written account. Enhanced ethnography is the second form and is an account of a single event, which exploits narrative devices drawn from novelists including the use of descriptive scenes with a clear point of view and the integration of dialogue. The author is embedded in the narrative either as a character or a third person narrator observing the events.

I have used this technique by writing vignettes derived from my own fieldwork which I present in this book to show how creative practitioners who participate in social media groups use photographs and video to extend and expand their practice into new forms. My aim in writing these vignettes was to explore and map some of the relationships actions such as video, poetry and photography have with contemporary everyday life with its affective and sensory dimensions, and how these might be understood within a broader context of creative practices with and for mobile and social media. I invited people in my personal Facebook and Twitter networks to participate in interviews between October 2013 and July 2014 to find out more about how they used social and mobile media for mundane as well as creative purposes. I was interested in the materialities, actions and processes and in the associations and practices that unfolded in the background, which were largely taken for granted. In addition to the interviews, I observed their use of social media on a daily basis.

I sought to evoke a sense of being there in these vignettes by placing the reader beside the central character through a multi-sensory and thick description to create emotional resonance and empathy. These accounts are not direct eyewitness recounts, rather they are "more-or-less what happened, but as a novelist might report it" (Humpheys and Watson 2009: 42). I also use a third person omniscient narrator as a narrative device or technique.

The third form identified by Humphreys and Watson (2009) is semi-fictionalized ethnography, which restructures events occurring within one or more observations into a single narrative using devices drawn from fictional writing. This really useful when using research material that is sensitive and where identities need to be protected. The truth of semi-fictionalized ethnographic accounts is pragmatic. They present accurate renditions of what happened where there are contrasting and contested versions of events. Humphreys and Watson present a semi-fictionalized example drawn from Humphreys's ethnographic study of corporate narratives. He found competing narratives and merged personalities to create a character called Charity to represent "those individuals who felt marginalized and somewhat at odds with most of the mainstream financial, marketing and IT specialists who were the core staff of the organization" (Humphrey and Watson 2009: 50). Humphreys writes an account of his fieldwork that is an amalgamation of events with composite characters that presents a short biography of Charity, one of the research subjects to show how management roles in large financial corporations can cause internal values conflicts for women from working class backgrounds. Creative practice researchers reflecting on their career experiences can use this technique to protect the identities of their informants.

The fourth and final form is fictionalized ethnography. In this form, the writing does not deal explicitly with theory. Rather it seeks to provide "an entertaining and edifying narrative" (Humphreys and Watson 2009: 42) that is based on lived ethnographic experiences. They provide an example of a worker who hoodwinks his bosses into thinking he is a good worker but impresses his co-workers with his mischief and deviance.

Ethnographic fiction is not new and may have some resemblances to creative non-fiction in that it relies on techniques and devices normally associated with fictional writing. Early ethnographic writing drew on fictionalizing techniques but then these were sacrificed for more empirical accounts. In the 1980s, anthropology like numerous other disciplines experienced the impacts of poststructuralism, which challenged notions like objective facts and a central truth. In the introduction to *Writing Cultures*, a landmark text for anthropology, Geertz declared, "Ethnographic writings can properly be called fictions in the sense of something made or fashioned" (1986: 6). Geertz famously wrote that it "is not clear just what 'faction', imaginative writing about real people in real places, exactly comes to beyond a clever coinage; but anthropology is going to have to find out if it is to continue as an intellectual force in contemporary culture" (Geertz 1988: 141). Clearly, anthropology and its signature ethnographic methodology have continued as an intellectual force. Indeed, ethnography has been adopted by other fields such as cultural studies, digital media and cultural geography as an important way of working.

The border between fiction and ethnography lies in the terrain between creative practice research and ethnography. I return to Narayan (1999) to help identify some of the writing actions and practices that distinguish ethnography from fiction. She argues that the differences between ethnography and fiction lie in the actual writing practices themselves:

What, then, distinguishes ethnography from fiction? While moving back and forth between ethnographic and fictional genres as I write of Kangra/"Triagarta" and reading the work of others engaged in similar travels, so far four practices have emerged as points of orientation. These are (1) disclosure of process, (2) generalization, (3) the uses of subjectivity, and (4) accountability. (139)

Jacobson and Larsen (2014) also explore this borderland from the perspective of cultural geographers using ethnography as a research methodology and state that the "purpose of writing ethnographic fiction is to *evoke* experience and meaning" from a narrative point of view. They focus on three literary techniques as a way to evoke place through creative writing: "verisimilitude (believability), kinesis (narrative movement), and scene-setting" (184). If we combine Narayan's (1999) four practices with the three techniques identified by Jacob and Larsen (2014) we have a powerful way to write about lived experience whether it be ethnographic fieldwork or our own creative practice. Experiences from fieldwork can be fictionalized and if this is done with careful attention to provide rich and thick descriptions to achieve verisimilitude with carefully plotted turning points a clear sense of what it was like to be there may be evoked to inform readers. If we inform our methodology with non-representational philosophical epistemologies and ontologies we can think about what lies between ethnography and creative practice research in new ways and push academic writing into new forms and genres.

Ethnographers have drawn on literary techniques that are the stuff of creative writing (Narayan 1999) to write up their fieldwork since the nineteenth century when cultural anthropology emerged as a discipline. At the same time, creative writers give their fictions verisimilitude through detailed descriptions of the nuances of cultural life, for example Chekhov's short stories and novellas are renowned for their faithful depictions of cultural life in Siberia and have been cited by Narayan (2012) in her book titled *Alive in the Writing: Crafting Ethnography in the Company of Chekhov*. W.G. Sebald's (1998) *Rings of Saturn* has become a point of reference for cultural and human geographers using fictional techniques to bring life to their writing about places (Jacobson and Larsen 2014: 184).

A Very Brief Conclusion

In a recent essay for the ethnography journal, *Savage Minds*, Roxanne Varsi, an ethnographer, film-maker and novelist writes

> Fiction, for me, like ethnography, has always melded with a deep desire to understand and explain the world around me. As an eight-year old in Iran I wrote stories to either escape or explain the Revolution that had turned my country into an Islamic Republic and had turned my single identity as a dorageh, or two-veined Iranian, into half-American, half-Iranian, forcing me to either choose one identity or to stay in-between. Writing helped me to make sense of the in-between, to make sense of my new life while holding on to the one that was already becoming a dream — unreal. (Varsi 2014, n.p.) (https://savageminds.org/2014/10/13/ethnographic-fiction-the-space-between/)

Varsi has found that combining ethnography and fiction together has allowed her to produce sharper accounts of cultural life in Iran that protected her informants and was true to her fieldwork exploring underground theatre in Iran. Her sentiments provide testimony for the beneficence of a mutual reciprocity between creative practice research and ethnography. Borderlands can be challenging for the academy but they are also spaces where new understandings and intuitions can be seeded and flourish. The fictional turn in ethnographic writing is concerned with aesthetics and poetics as well as with symbolic and sociological structures that govern social activities and everyday practices in specific cultural groups. I conclude this chapter with the observation that there is much to be gained from allowing creative practice research and ethnography to mingle and form attachments. There is certainly chemistry between them. In the next chapter, I explore how the remediation of old forms through mobile media and the extreme accessibility of smartphones seeds new visual vernaculars through my ethnographic research with self-declared creative artists using mobile and social media.

REFERENCES

Anderson, B., and Harrison, P. 2010. The promise of non-representational theories. In *Taking Place: Non-Representational Theories and Geography*, ed. Anderson B and Harrison P, 1–36. London: Ashgate.

Austin, John, and Langshaw. 1962. *How to Do Things with Words*. Oxford: Oxford University Press.

Baszanger, I., and N. Dodier. 2004. Ethnography: Relating the Part to the Whole. In *Qualitative Research Theory, Method and Practice*, 2nd ed, ed. D. Silverman, 9–34. London: Sage.

Batty, Craig and Berry, Marsha. 2015. Constellations and Connections: The Playful Space of the Creative Practice Research Degree. *Journal of Media Practice* 16 (3) 181–194. http://dx.doi.org/10.1080/14682753.2015.111 6753.

Batty, Craig, Sung-Ju, Suya Lee, Sawtell, Louise, Sculley, Stephen, and Taylor, Stayci. 2015. 'Rewriting, Remaking and Rediscovering Screenwriting Practice: When the Screenwriter Becomes Practitioner-researcher'. *Writing the Ghost Train: Refereed conference papers of the 20th Annual AAWP Conference*, 2015.

Berry, Marsha. 2016. Out in the Open: Locating New Vernacular Practices with Smartphones. *Studies in Australasian Cinema* 10 (1) 53–64. http://www.tandfonline.com/doi/abs/10.1080/17503175.2015.1084173.

Berry, Marsha. 2017. Ethnography and Screen Production. In (forthcoming).

Brabazon, T., and Z. Dagli. 2010. Putting the Doctorate into Practice, and the Practice into Doctorates: Creating a New Space for Quality Scholarship Through Creativity. *Nebula* 7 (1–2): 23–43.

Bolton, Gillie. 2010. *Reflective Practice: Writing & Professional Development*. London: Sage

Clifford, J. 1986. 'Introduction: Partial Truths'. In *Writing Culture: The Poetics and Politics of Ethnography*, ed. James Clifford and George E. Marcus. Berkeley: University of California Press.

Emerson, R.M., R.I. Fretz, and L.L. Shaw. 1995. *Writing Ethnographic Fieldnotes*. Chicago: University of Chicago Press.

Erving, Goffman. 1989. "On Fieldwork". *Journal of Contemporary Ethnography* 18 (2): 123–132.

Geertz, Clifford. 1988. *Works and Lives: The Anthropologist as Author*. Stanford: Stanford University Press.

Haseman, Brad. 2006. A Manifesto for Performative Research. *Media International Australia incorporating Culture and Policy, theme issue "Practice-led.Research"* (no. 118): 98–106.

Hjorth, Larissa, and Sarah Pink. 2014. New Visualities and the Digital Wayfarer: Reconceptualizing Camera Phone Photography and Locative Media. *Mobile Media & Communication* 2 (1): 40–57.

Husserl, Edmund. 1954 (1970). *The Crisis of European Sciences and Transcendental Phenomenology*. Evanston: Northwestern University Press. http://s3.amazonaws.com/academia.edu.documents/34337886/ Husserlcrisis.pdf?AWSAccessKeyId=AKIAIWOWYYGZ2Y53UL3A&Expire s=1488518588&Signature=q4ETJQ95ohYGnfSm8wHs%2BFXGctU%3D &response-content-disposition=inline%3B%20filename%3DThe_Crisis_of_ European_Sciences_and_Tran.pdf.

Humphreys, M., and T. Watson. 2009. Ethnographic Practices: From 'Writing Up Ethnographic Research' to 'Ethnographic Writing.' In *Organizational Ethnography: Studying the Complexity of Everyday Life*, ed. Sierk Ybema, Dvora Yanow, Harry Wels, and Frans H. Kamsteeg. London: Sage. Also available at http://www.corwin.com/upm-data/26764_03_Ybema_Ch_02.pdf.

Ingold, Tim. 2011. *Being Alive: Essays on Movement, Knowledge and Description*. Oxon: Routledge.

Ingold, Tim. 2015. Foreword. In *Non-representational Methodologies: Re-envisioning Research*, ed. Phillip Vannini, pp. vii–viii. New York and London: Routledge.

Jacobson, Marc, and Soren Larsen. 2014. Ethnographic Fiction for Writing and Research in Cultural Geography. *Journal of Cultural Geography* 31 (2): 179–193.

Kerrigan, Susan, Leo Berkeley, Sean Maher, Michael Sergi, and Alison Wotherspoon. 2015. Screen Production Enquiry: A Study of Five Australian Doctorates. *Studies in Australasian Cinema* 9 (2): 93–109.

Lorimer, H. 2005. Cultural Geography: The Busyness of Being "More Than Representational". *Progress in Human Geography* 29 (1): 83–94.

Maynard, Kent. 2002. An "Imagination of Order": The Suspicion of Structure in Anthropology and Poetry. *The Antioch Review* 60: 2, Angels & Devils, 220–243.

Narayan, Kirin. 1999. Ethnography and Fiction: Where is the Border?. *Anthropology and Humanism* 24 (2): 134–147.

Narayan, Kirin. 2012. *Alive in the Writing: Crafting Ethnography in the Company of Chekhov*. Chicago: University of Chicago Press.

Pink, Sarah. 2001. *Doing Visual Ethnography*. London, England: Sage.

Pink, Sarah. 2009. *Doing Sensory Ethnography*. London: Sage.

Schechner, Richard. 2002. *Performance Studies: An Introduction*. London: Routledge.

Sebald, Winfried Georg. 1998. *The Rings of Saturn*. New York: New Directions.

Sullivan, Graeme. 2009. Making Space: The Purpose and Place of Practice-led Research. In *Practice-led research, research-led practice in the creative arts*, ed. H. Smith and R. Dean, 41–65.

Thrift, N. 2008. *Non-representational Theory: Space, Politics, Affect*. London: Routledge.

Vannini, Phillip. 2015a. Non-representational Ethnography: New Ways of Animating Lifeworlds. *Cultural Geographies* 22 (2): 317–327.

Vannini, Phillip. 2015b. Introduction. In *Non-representational Methodologies: Re-envisioning Research*, ed. Phillip Vannini, 1–19. New York and London: Routledge.

Varsi, Roxanne. 2014. 'Ethnographic Fiction: The Space Between'. *Savage Minds: Notes and Queries in Anthropology*. https://savageminds.org/tag/roxanne-varzi/. Accessed 23 Aug 2017.

CHAPTER 2

Playing with Visual Vernaculars

INTRODUCTION

John is sitting by the window on the train. He's heading into work in Melbourne's CBD. It's mid December and the day promises to be hot. He checks Facebook to pass the time. He sees a photo of the first snow of the year taken by a friend earlier. His friend lives in Holland. The photo is carefully framed with a vignette effect. It has a melancholy air and the caption "Time to hibernate". John 'likes' the photo and takes a photo of out of the train window, adds an ironic comment about the hot north wind that is currently blowing in Melbourne to the photo and posts it to the comments section of his friend's post. His friend replies, "send it here, but with my luck it'll come via the Siberian route". John finds and posts a picture of Grumpy cat. He grins, anticipating a feisty response from his friend.

Social media has reignited John's old love for writing poetry and taking photographs and now he has an audience. It has given him a reason to write poems again and to take photographs. He found it pointless before because his friends and family treated it as one of his indulgent hobbies. Now he has found a network of people who share their poems and images on a daily basis and give each other encouragement to stretch their wings. He gladly jumps into social media streams every day to share his own work and to appreciate the work of his newly found fellow travellers. People perform and enact their creative impulses in their social media timelines through new vernaculars and cultural memes such as lol cats. Social media is floating world of evolving nuanced cultural

© The Author(s) 2017
M. Berry, *Creating with Mobile Media*,
DOI 10.1007/978-3-319-65316-7_2

meaning systems that sits in the background where we spend time with family, friends and fellow travellers who share our interests. Social media spaces have connections to physical environments as well. They transcend specific local time zones, seasons and weather so that people may be in a common social media space synchronously yet be experiencing completely different local physical conditions. Social media collapses weather and time zones so these can longer be taken for granted when we communicate with online interlocutors. My question in this chapter is: How do people play with emergent visual vernaculars to enact their creative practice?

In previous work (Berry 2015), I have identified how visual vernaculars have developed as a way of talking about these differences using images as a form of shorthand. The vignette at the start of this chapter is an illustration of how social media posts use photographs as an idiomatic shorthand to evoke contrasting sensory perspectives of the present moment in two geographic locations with different seasons. To better understand the nuances of how new visual vernaculars are improvised and enacted, I draw on Ingold's (2007, 2008) concept of a zone of entanglement, which allows me to re-explore the messy connections between people and social media evident in the opening vignette and unpack how creative practices are intertwined with other daily and routine activities. I use his concept as a theoretical framing device think about emergent vernacular film-making and photography in non-representational ways and interpret the some of the implications of the ubiquitous presence of smartphones for creative practice.

Opportunities for image making (Schleser et al. 2013; Keep and Berry 2013) are embedded in the background of everyday routines because of the affordances of smartphones. I wanted to explore and map some of the relationships creative and evocative expressions such as video, poetry and photographs have with contemporary everyday life with its affective and sensory dimensions and how these might be understood within a broader context of emerging visualities and social media. My focus quite deliberately includes what is in the background, on the taken for granted and habitual aspects so that I can examine thoughts and creative practices in action using a non-representational lens (Thrift 2008; Vannini 2015; Ingold 2010). I invited people in my personal Facebook and Twitter networks to participate in interviews about their use of social media at the start of October 2013 to find out more about how they used social and mobile media for mundane as well as creative purposes.

According to Postill and Pink (2012), the distinction between online and offline is problematic so they advocated that to understand the impacts of social media, one really ought to look at how "research participants navigate their wider social, material and technological worlds" (Postill and Pink 2012, 123). So, to address the importance of being able to gain insights into the entanglements of the online and offline, I asked my participants questions about where social media sat in relation to the totality of their social relations as well as their creative practices.

Within a fortnight, I had received sixteen responses from self-identified creative practitioners from Australia, USA, UK, Pakistan, Korea and the Netherlands. I also had their permission to observe their public interactions with others in social media provided I kept their identities anonymous. Collectively, my respondents all reported daily routines with different social media platforms, which were embedded alongside other activities including their creative practice and social lives. They frequently posted images connected to where they were and referred to what they were doing or thinking in the present moment. They also posted numerous creative expressions such as photographs and short form poetry to special interest groups for artists, writers and photographers.

This chapter contains extracts and vignettes derived from the content of the interviews. The extracts from interviews are presented verbatim. To compose the vignettes in this chapter, I used a writing technique, commonly used in anthropology, called enhanced ethnography (Humphreys and Watson 2009) to protect the identities of my participants. I provide a more expansive discussion of this way of working in Chap. 1. Drawing on my own practice as a creative writer, I fictionalized the characters or protagonists for the vignettes and used my field notes and interviews for the patterns of events. In this way, I have used techniques more commonly associated with fiction writing forms interweaving "novel-like detailed descriptions of natural sounding events" (Humphreys and Watson 2009, 48) to present the information gathered from my informants about routines with social media in an accurate yet evocative way.

The extracts and vignettes show that zones of entanglement between online and offline include improvisations by artists and poets which create new vernaculars. These zones also include a strong sense of audience, may rely on phatic communication and rapport and provide spaces for the enactment of new and emplaced visualities. This chapter is organized into four sections. The first section discusses the emergence of new vernaculars within the context of participatory cultures; the second explores the importance of rapport and audience for creative practice and how the use of social media facilitates this within creative practice networks; the third section focuses on how visualities are enacted within everyday routines and figured in the emergence of new vernaculars; and the final section returns to zones of entanglement and the unfolding of creative practices to provide some concluding remarks about how new vernaculars are enacted and improvised.

USING NEW VERNACULARS FOR PARTICIPATION

Creative vernaculars are a series of creative practices located in the everyday that now include vernacular photography and film-making. They may be figurative as well as literal in their meaning. These vernacular practices predate networked mobile media but have taken on new forms in social media, whereby the digital objects that are produced are shared with a wider audience than their analogue counterparts and participate in broader local and global conversations. For example, demotic forms such as family photo albums, holiday movies and scrapbooks can be shared and passed on through social media and these participate in public conversations. In the context of smartphone cameras and social media, these practices highlight the popularity of cultural participation and present new opportunities for creative practice.

Burgess (2008) describes vernacular creativity as "both an ideal and an heuristic device, to describe and illuminate creative practices that emerge from highly particular and non-elite social contexts and communicative conventions" (p. 215). She situates her discussion from a cultural studies perspective and proposes that "the term signifies what Chris Atton calls 'the capacity to reduce cultural distance' between the conditions of cultural production and the everyday experiences from which they are derived and to which they return (2001)" (p. 216). There are numerous Facebook groups that encourage non-elite participation. For example, the London-based public Facebook group called Ghostsigns[1] founded by Sam Roberts has reduced both cultural and geographic distance and relies on participation for its content. The site is an informal archive of hand-painted advertising on the brickwork of buildings. The main photographic images on the timeline are from contributors all over the UK and are examples of vernacular creativity. The conversations through the comments on the timeline reveal how people have become inspired to find and share ghost signs in their own local areas by Sam Roberts's initial photographs. The *Mirror*[2] has picked up on the increasing popularity of ghost signs.

Participatory cultures, such as the Facebook group Ghostsigns, have grown around social media platforms and are facilitated by smartphones. The lines between producers and consumers of content have become truly blurred, and this is a key factor driving the emergence of new vernaculars. People improvise ways of sharing by remixing and remediating images (Bolter and Grushin 1999) with the look and feel of old and new media to communicate aspects of their lives and creative undertakings to create new idiomatic expressions. Innovative forms are being used within spaces created by social media where new creative vernaculars (Burgess 2006) are burgeoning, and now we have the ability to observe the unfolding of people's creative processes through social media networks. We can trace and observe their interactions with other people and their social networks.

Another public group on Facebook called Haiku provides eloquent and evocative examples of how visual images have been added to an ancient form of poetry to remediate it, and how in turn haiku now has vernacular dimensions. One post in this group's timeline in April 2015 caught my attention. It was an image taken with a smartphone of bird poop on a car's side mirror which was overlaid with this haiku: "blue Robin / left his presence / behind".[3] The comments left on the post were humorous, informal and supportive. One read, "Lovely, it must be Spring—lol (smiley face)", the author of the post responded "Indeed! (smiley face)"; another said "glad it was not Christopher Robin".

Both of these Facebook groups have opened up new places for people like John, from the opening vignette where they can express themselves and find new friends who inspire their creative endeavours. Social media started out as a subjunctive play space for John. He started tentatively posting haikus in 2010 on Twitter in response to challenge words. He was pleasantly surprised that people responded favourably to his offerings. He opened a Facebook account when his friends on Twitter encouraged him to join them there because there was more scope to post longer poems and have more detailed conversations because there was no 140-character limit. Again he found a wealth of contacts that shared his love for poetry and photography, so he was inspired to start experimenting with smartphone photography and video applications. Building rapport with his audience also means much to John. In the next section, I discuss how people seek to create rapport through visual vernaculars.

BUILDING RAPPORT IN SENSORY WAYS

Sharing objects such as photographs and video has become a part of our daily routines (Pink and Hjorth 2012). We are out and about in all kinds of weather going about our routine business such as commuting to work, eating, playing with our pets, and we can too share these with our social circles engaging in a networked sociality (Wittel 2001). We communicate many moments of our embodied and phenomenological existence to bring others into our lives through social media. The photographs taken by John and his friend are not only representational in terms of the weather conditions but also performative in function. They serve to mesh the kinetic and phenomenological aspects of two separate lifeworlds and paths together in one social space where both interlocutors are synchronously co-present (Ito 2006), even though John is in Melbourne enjoying a hot day in summer and his interlocutor is complaining of a cold winter in Holland. The photographs have vitality, and it may be said that they participate in an ethos of animation (Vannini 2015).

It has become very easy to share an image with others who are synchronously co-present online with smartphones. And if they are not synchronously co-present, they can see the posts later. The mendacities of life and associated thoughts are openly posted including the details of where people are walking, what they are eating and whom they are with. Photographs and videos are clear ways of creating empathy for our locomotive, affective and embodied experiences in the age of ubiquitous computing and social media. Visual images can readily communicate a sense of what it is like to be here and now in the physical world in an animated, evocative and visceral way. Jenny, a grandmother in her sixties and who lives alone in Wales, is a case in point of the urge to build rapport in sensory ways through mobile social media. She has been a practising artist for many years. Her intention to share multisensory aspects of her life is evident in the vignette below drawn from her response to a question about the kinds of photographs she shares and what motivates her to share them. Her smartphone has become integral to her creative practice and allows her to participate in an ethos of animation (Vannini 2015).

Frogs, Wasps and an Abandoned Liner

It was a mild early summer morning. Jenny was sitting by the water irises by her pond having her breakfast of tea and toast. The frogs were sunning themselves on the stones around her garden pond. She recollected the mixed reactions she'd had last night sparked by a picture she'd taken of a wasp. Photos were a good way of starting conversations and making friends online she thought.

She stretched, enjoying the soft sun on her face. She took a picture of a frog on a stone with her phone and shared it to Instagram, Facebook and Twitter saying how much she loves her garden frogs. A message came back saying "kiss him and he might turn into a prince". She grinned and responded, "I don't want no princes, I want a man who can do the heavy work 'round here". The bantering continued while Jenny drank her tea. This gave her an idea for a series of photographs called A Marriage of Frogs and Wasps …

She mulled this idea over as she drove to her local garden center. When she arrived she checked Twitter on her smartphone and saw that over a dozen people had retweeted her wasp and frog photos. She look a photo of a dragon statue in the garden center, inserted a text box using a photo app saying "Thank you to all for retweeting my photos" and posted it to Twitter. She bought the two bags of potting mix she had come for.

On her way home, she decided to stop by an abandoned ocean liner. She took lots of photos cycling through her smartphone apps playing with the filter effects. The gathering clouds and rising wind added mystery to the scene. She played with the idea that the liner was haunted. She felt she had enough now for a show at the local gallery where she had regular exhibitions. She posted several of these pictures on Twitter, Pinterest and Instagram. One image was just posted to the timeline without the use of any filter effects, others she altered and added a text overlay with poetry. She waited anxiously for reactions from her online circle of acquaintances. She's planning to use the feedback in her application for the gallery space. She was hoping they'd be open to a show featuring images she'd taken with her smartphone.

"Why not," she thought, "there was a show featuring mobile phone photos at one of the major galleries in London recently."

Jenny's smartphone creative practice and social media practices show how "life takes shape and gains expression in shared experiences, everyday routines … unexceptional interactions and affective dispositions" (Lorimer 2005). Banter has become a feature of her art-making actions and process. There are numerous small actions that make up her practice—some of them quite playful and for her it is also a way of socializing and building rapport in social media spaces.

Geographical distance from loved ones and the need to stay connected to familiar social networks also can become an instigating force for sharing vernacular and sometimes banal photographs through social media on a regular basis. Min is a Korean international student undertaking doctoral studies in fine art who currently resides in Melbourne and relies on social media to share the details of her life with her friends and family back in Korea. She told me that she uses Facebook, Twitter and Kakao Story (Korean SNS). When asked to describe how using social media fits into her days, she says:

> It is like a daily routine, same as having meal, I reckon. When I wake up in the morning, I check mine and my friends' Kakao story's photos and comments first. Then I take photos of mundane things such as my dog, breakfast, and weather…and upload them to my Kakao Story. So I can talk with my friends and family without seeing each other at the cafe. But I am so happy because they know my life in Melbourne and I know what they're doing in Seoul everyday.

Min was a special case in my study in that she is a transient migrant and I pick up on this later. My concern here is that building and maintaining relationships were important to all of the participants in my digital ethnography and many relied on creative expressions that also functioned as phatic communication to cement ongoing relationships. For now, I wish to discuss the emergence of using vernacular images as a form of social glue. Numerous studies explore phatic communication in social media (see, e.g. Radovanovic and Ragnedda 2012; Marwick and boyd 2010; Miller 2008). Phatic communication is "a type of speech in which ties of union are created by a mere exchange of words" (Malinowski 1923: 151). Its main purpose is to create a rapport between interlocutors. Roman Jacobson identified a phatic function for language where the objective is to maintain social ties and relationships. He refers to small talk as an example. Opinions on the value of phatic communication in social media networks are varied. According to Radovanovic and Ragnedda (2012), "the relevance of the phatic function of microposts is emerging as a form of online intimacy and of social connections in social networks " (p. 13). The phatic aspects of social media figure prominently in Min's days and use of visual vernacular images in her posts do create a sense of intimacy with her family and friends in Korea. They perform as a form of affectionate small talk to reduce a sense of distance.

Nonetheless, some new media researchers denounce the emergence of phatic communication in online ecologies as being nihilistic and without substantial content. For example, Vincent Miller (2008, 398) asserts that we "are seeing how in many ways the internet has become as much about interaction with others as it has about accessing information". He claims "communication has been subordinated to the role of the simple maintenance of ever expanding networks and the notion of a connected presence" (p. 398) and that "communication without content has taken precedence" (p. 398) and concludes that we are moving away from "communities, narratives, substantive communication, and towards networks, databases and phatic communion" (p. 399) where we are obliged to remain connected to others through posts and yet "remaining fairly oblivious as to the consumption (and production) of information" (p. 399).

There is no denying that people do spend a substantial amount of time in social media networks on phatic communication; however, a more fine-grained approach to rapport building and phatic communication is needed to better understand the complexities of sharing and interacting in these environments. I have found concrete evidence to counter Miller's (2008) more abstract conceptual position that phatic communication is effectively oblivious to substance. Min's response indicates that she has not moved away from communities, narratives and substantive communication; indeed, the phatic and rapport building aspects of her routine sharing of updates with friends and family in Korea serve to reinforce her narrative of living as an international doctoral candidate in Melbourne. Min is not an isolated case. Numerous studies attest to the importance of phatic communication as a rapport building tactic in the online interactions of transient migrants. For example, Panagakos and Horst (2006) argue that migrants often use communication technology to stay connected, communicate and create co-presence.

A sense of rapport with both her imagined and known audience is also an intention of Jenny's creative improvisations as the vignette above shows, her use of social media and smartphones do include phatic dimensions to stay connected but she is certainly not oblivious to content either. Contexts, where the exchange of creative expressions is the norm may be created, maintained and extended to include sensory and affective aspects through phatic communication. What my ethnography reveals is that "much like writers, social media participants imagine an audience and tailor their online writing to match" (Marwick and boyd 2010, 15). Building rapport and sharing small talk are definitely intertwined with audience awareness and improvisation for my participants.

Another case illustrative of the desire to build rapport with imagined and known audiences that I derive from my ethnography is Paul. He is another like John who has been inspired through his social media interactions to join the ranks of creative artists. He is a retired psychotherapist who has taken up creative writing and photo-media art since his retirement. He has a keen sense of his potential audience and is consciously aware of the content he produces and shares as a tactic "to get his work out there". He lives in the UK. His images have appeared as covers on his friends' poetry volumes and he is a published author. He is self-deprecating about his photography and sees it as a hobby he enjoys immensely, whereas he takes his writing seriously—he sees himself as a writer rather than a visual artist. When I asked him how he uses his smartphone camera and social media, he responded:

> I photograph local scenery/landscapes, family & friends. I also create photoart: compilations of images on a certain theme or 'straight' images that I've changed & embellished using Apple iPhone photo apps. Iphoneography has become a valued hobby, to the point where I've almost abandoned my DSLR camera as I find it easier & much more fun to take & edit photos on my iPhone. It's a hobby you can pick up & put down as & when time/space dictates ... I find it very enjoyable & fulfilling. I also enjoy having my ego stroked when people comment positively on my photos, either on Facebook or Instagram.

Rapport and phatic communication is an important precondition for the emergence and circulation of creative visual vernaculars. In the next section, I consider how people enact visualities and visual vernaculars with mobile media and how such actions and enactments reflect and shape creative impulses.

Enacting Visuality Online and Offline

In order to discuss how visualities and visual vernaculars in play out in mobile media, I first turn my attention to situate co-presence. Digital or networked co-presence (Okabe and Ito 2006) is a feature of our daily encounters with geographic places. The notion of a networked co-presence presupposes the centrality of co-presence to social interaction. In his book *Behavior in public places* (1963), Goffman analysed contexts of co-presence. According to Goffman, an awareness of presence is essential for "persons must sense that they are close enough to be perceived in whatever they are doing, including their experiencing of others, and close enough to be perceived in this sensing of being perceived" (Goffman 1963: 17). All face-to-face interactions and encounters have the physical co-presence of interlocutors as given in Goffman's theory. He identified gatherings as two or more individuals who are in each other's immediate presence and situations as "the full spatial environment anywhere within which an entering person becomes a member of the gathering that is (or does then become) present" (Goffman 1963: 18). But in online environments, physical co-presence is not given and spaces are fluid and dynamic. I propose that contexts for co-presence have expanded and now include social media as situations where networked co-present gatherings take place and that these situations are now ever-present in the background of everyday life.

Furthermore, new kinds of co-presence are emerging where online and offline cartographies have become meshed together. Pink and Hjorth (2012) theorize such social phenomena as "a shift from networked visuality to emplaced visuality and sociality" (p. 145). Anyone can easily observe networked and emplaced visualities on his or her social media timelines where people post, share and "like" each other's visual images. Networked co-present situations and gatherings do provide an impetus for the evolution of new visualities and creative vernaculars. The ease of enacting visualities with the current smooth and almost immediate connections between online and offline with smartphones was arguably "unimaginable even a decade ago" (Vickers 2013, 139). Social media timelines are filled with examples of spaces "where many things gather, not just deliberative humans, but a diverse array of actors and forces, some of which we know about, some not, and some just on the edge of awareness" (Anderson and Harrison 2010, 10). Furthermore, such spaces or ecologies are a part of a liquid modernity that Bauman (2003) theorized as being one where "change is *the only* permanence and uncertainty *the only* certainty" (Bauman 2012, viii–ix). Creative visual vernaculars are adaptive responses to the dynamics and ambiguities of mobile media ecologies.

Social and mobile media assemblages and affordances are encouraging many to participate in creative practice networks (Berry 2011), which in turn are creating constellations of narratives with their own rhetoric about the use and place of technology. To show how emergent visualities are enacted on an everyday basis, I refer to the case of Joseph (Berry 2015). Joseph is a published poet in his homeland Holland who also works as a project engineer as a consultant. In his interview with me in 2013, he reported that he found co-dependence between his inspiration and his connections on social media, especially Twitter.

A Routine day

A dog nuzzles his sleeping master. The sky is streaked with pink and orange. Joseph wakes and stretches his arms over his head. He wonders what kind of response he has received from the project finance manager. The project is going well, Joseph thinks to himself and smiles. He remembers how he dreamed of this day as a young man, a day when he could work at home on design projects and chat to his diaspora family each day by video so he could see their faces.

The boiled egg makes a most satisfying crack this morning. Joseph opens the lid of his laptop whilst absentmindedly sipping his coffee and dipping toast into the runny centre of his boiled egg. Twitter and Facebook sit side by side on two screens and smiles at the latest poetry. He loves the way Tweetdeck lets him look at several hash tags at once. He writes a short haiku about his boiled egg likening it a new dawn. Not convinced by his metaphor he posts it anyway. Within moments it is retweeted.

His camera, lying on the desk beckons, the videos of his dogs on their evening walk yesterday worked out quite well. People enjoy the antics of his two springer spaniels with their goofy, innocent faces. Time to bring a little light into the world, he thinks and uploads the best sequence to Facebook. He nods with satisfaction. Yes, the sequence does justice to the melancholic clouds in the late afternoon light and the dogs make a wonderful counterpoint. He posts a haiku about clouds growing older in Twitter and pastes it to a Facebook group for poets and artists called Small Stones. Tempted to wait for comments and likes, he tousles his dog's ears, kisses their noses and tells them that it's time for work. Twitter and Facebook now sit in the background.

He opens a second web browser and logs into his professional email and the enterprise system and share drives. No dramas there. He responds to routine questions. He sends the revised specifications to the team. Joseph flicks back to Facebook and sees that ten people have liked yesterday's photos and video of his dogs and five have left comments. He sees a direct message from his cousin in the UK. He can see she is still online so they chat about the upcoming Easter break and his forthcoming trip to London. As he waits for her responses, he is checking the various haiku hash tags on Twitter. He is hoping there is a challenge to use a particular word but alas, nothing today. Indeed, there hasn't been much action for the past month. He suspects there will be more action over in the haiku groups on Facebook so he opens a haiku closed group page. His suspicions are confirmed and the funny thing is that all his old contacts from Twitter seem to have migrated to Facebook. Why is that, he wonders?

His work email beeps insistently, interrupting his reverie. He answers it, shakes his head at the questions with obvious answers and opens the enterprise database. He stands up stretches, goes to the kitchen, boils the kettle and makes coffee. The smell gives him strength. It really is time to get back to work. He steadfastly responds to the emails demanding attention. He hopes they have a good filing system in the organization because some of the email chains are over twenty messages long.

After a couple of hours, the dogs come into his study; carrying food bowls in their mouths. Joseph's stomach is complaining too. He feeds the dogs and sets out with them both for his favourite local café for lunch. They walk back through the fields. He stops to take a picture of the vista and posts it to Instagram and Facebook straight away. He adds a caption to let people know he is out walking with his dogs. Bruno has found an enormous stick and is struggling to carry it. Joseph laughs at the sight. He posts a 15s video to Instagram, Facebook and Twitter simultaneously with a humorous caption. Others will share his mirth, for sure. And sure enough, a friend from Yorkshire sends back a smiley face. He continues his walk home through the fields with the dogs that insist on smelling every tree. (Berry 2015, 57–58)

The ease of switching between online and offline provided by smartphones issues us with a challenge to rethink video and photography practices. What is evident in Joseph's story is the degree to which creative vernaculars such as writing haiku and taking photographs are a part of mundane routines such as taking one's dogs for a walk. These creative practices sit in the background of everything Joseph does, just waiting for the right impetus. Photography in social media environments has expanded exponentially incorporate what Goggin (2006) identified as a "demotic turn" (Goggin 2006, 146). This demotic turn is evident in Joseph's use of creative vernaculars. Indeed, since Goggin wrote about the demotic turn in photography in 2006, photography has changed even further because of the extreme accessibility of smartphones and can consider commonplace. Online and offline worlds are essentially mingled so that it is possible for Joseph to go walking alone with a dog, yet be accompanied by digitally co-present others (Hjorth and Pink 2014). (I pick up on this notion more in Chap. 4.) He can take photographs, use filters to create visual expressions, share these almost immediately and receive a response.

In 2009, Jenkins proposed that a new form of culture with participation as its hallmark was emerging through the use of social media and Web 2.0 technologies. These participatory cultures have indeed arisen where anyone with a smartphone can participate and, in turn, shape creative vernaculars. In addition, participatory cultures are alive and evolving through the demotic turn and associated vernaculars in similar ways to language usage. Emergent visualities and socialities (Pink and Hjorth 2012) do encourage people to participate actively with new and remediated (Bolter and Grushin 1999) forms of video and photography, in complex media ecologies where the lines between online and offline are blurred and have become increasingly meaningless. The respondents in my ethnography are a part of these evolving participatory cultures and media ecologies that have volume and immediacy as their hallmarks. They all indicated that social media participation was a potential source of distraction as well as inspiration for them. As my respondent Jane, a single mother from the UK, put it, "I find social media really enjoyable so I purposely try to apply some boundaries to prevent me from spending too much time on it at the expense of other things". By day, she works as a freelance professional writer but she aspires to become a journalist.

> Jane looks rushed. The deadline for the technical report she is editing draws closer. She blows a raspberry at the clunky language. She sits at her computer, looks at Facebook on her smart phone. She really should check LinkedIn to see if there are any leads for jobs. Freelance work is great but she is worried about her rent. Once the report is done she will be unemployed. The kids need new shoes too. There are a couple of promising prospects. She makes a note of them and relaxes. One of them will come through for sure. She tweets out a link to her writing portfolio site and checks her number of followers. She has broken through the 3000 mark. Her portfolio blog has over 60,000 hits. That will look good on her CV.

> She returns to Facebook. Her close friend Lyn has posted a new poem. She shares it and then reads it slowly, savoring the images - lyrical as ever.

There's a link to a video of a cat drinking straight from a tap. Jane sniggers and posts a comment. Others join in. Cats are always a good talking point. She writes a short poem about cats and posts it to her poetry-writing group. There's a post there from Jack. The anthology they self-published has raised over $400 for UNICEF so far.

She returns to copy editing the report thinking social networking is way too much fun. Perhaps it's time to use the application on her Smartphone she downloaded yesterday that blocks her access to the Internet for a few hours so she can meet her deadline.

Jane's case shows how that boundaries between online and offline have become blurred and that a binary approach to conceptualizing mobile and social media spaces in opposition to geographic places has become redundant. However, if we look at Jane's action through a lens drawn from non-representational theory (Thrift 2008), we can begin to appreciate how visual vernaculars are emerging from ordinary everyday practices because of the use of smartphone assemblages. They are also a response to the rapid pace of change and the ambiguities and misunderstanding that arise all too easily in social media.

CONCLUDING REMARKS: UNFOLDING CREATIVE PRACTICES

Digital and mobile media is part of our material culture and is woven into the fabric of every life. Horst and Miller (2012) claim that

> Being human is a cultural and normative concept. We may employ technologies to shape our conceptualization of what it means to be human, but it is our definition of being human that mediates the technology, not the other way around. (Horst and Miller 2012, 108)

Through unfolding the actions that make up the creative practices of my participants, we can see that their working and tacit conceptualizations of what it means to be human are both messy and dynamic. They adopt and adapt creative visual vernaculars to socialize and to express themselves. The emerging phenomena associated with socially networked media provide opportunities to rethink how creative vernaculars arise, and why the mundane aspects of everyday life hold such compelling appeal. In the introduction of this chapter, I suggested that we also have unprecedented access to people's thoughts, feelings and creative processes through social media and smartphones. I have put forward clear evidence of the seamless yet messy interrelationships between my participants' embodied daily routines and their social interactions using networked communication and smartphones to support my claims.

Phatic communication is a significant feature of global social media landscapes where contexts have collapsed (Marwick and boyd 2010). New types of visual vernaculars are improvised and enacted in these landscapes. These vernaculars traverse global time zones to make bridges between local cultural contexts. John was killing time scrolling through his Facebook timeline on a hot morning in December in Melbourne. A photograph posted by a friend in the UK captured his attention so he responded in kind with visual images to bridge the distance between them. Jenny turned to the intertextuality of fairy tales and poetic aesthetics to communicate the embodied and affective dimensions of her life and used her posts to improvise creative outputs as drafts for her concepts for future exhibitions. Min photographed the stuff of her daily life to share with her friends and family in her home country so as to reduce the geographic distance between them. The actions of the protagonists in my vignettes all display a finely tuned awareness of known and imagined interlocutors and audience within their social media circles.

The sense of a networked and digital co-presence in social media situations is also a sensibility shared by my participants. The story of Joseph shows that my participants were not only conscious of potential audiences for their creative outputs but also assumed their networked co-presence. Joseph sensed that others would perceive him if he posted an update to his Instagram and Facebook timelines about walking with his dogs. The thought of having networked co-present followers and friends there with him while he was walking alone with his dogs inspired him to make a humorous video clip of one of his dogs lugging a huge stick. His assumption about being with networked co-present acquaintances on his walk was borne out as correct when he received a response from his friend who lives in Yorkshire.

The participatory cultures found in social media have provided a space for drifting, jamming and avoiding the task at hand for my respondents. For example, Jane is easily distracted from looming deadlines by her social media. At the same time, these floating worlds enabled by smartphones and new visual vernaculars can be a catalyst for creative pursuits such as poetry. For Jane, social media is a space to get involved collaborative creative projects such as poetry anthologies because "wireless, mobile and ubiquitous technologies are a portal to new modes of experience, thanks to which a user can be part of the bigger picture in the mediascape" (Schleser 2013, 94). Being part of a broader mediascape through having a significant online presence is important for Jane's portfolio career as a freelance writer as well as for her creative practice as a poet.

To sum up, being human and our relationship with mobile media technologies are a messy entanglement of the embodied, phenomenological and the mediated. We communicate across time zones, geographic locations, seasonal weather conditions as well as various and diverse cultural contexts. Ingold's theory of zones of entanglement does provide a novel way to think about these interrelationships in a way that acknowledges their complex and dynamic nature. Smartphones have become taken for granted and have given rise to the emergence of new creative vernaculars. They sit in the background providing a means to capture and trace our paths, thoughts, feelings and inspirations as we go about our everyday business. In short, smartphones and social media expand the conditions of possibility for creative practice. In the next chapter, I explore the selfie phenomenon and draw links between selfies, emplaced visualities and the emergence of creative visual vernaculars.

NOTES

1. https://www.facebook.com/ghostsigns?fref=ts.
2. http://www.mirror.co.uk/news/uk-news/ghost-signs-secrets-fascinating-adverts-5462049.
3. https://www.facebook.com/photo.php?fbid=980745885292932&set=gm.10152782810367547&type=1&theater.

REFERENCES

Anderson B., and P. Harrison. 2010. The promise of non-representational theories. In *Taking place: Non-representational theories and geography*, eds. B. Anderson, and P. Harrison, 1–36. London: Ashgate.

Atton, C. 2001. The Mundane and its Reproduction in Alternative Media, *Journal of Mundane Behavior* 3 (1), n.p.

Bauman, Zigmund. 2003. *Liquid Love: On the Frailty of Human Bonds.* Cambridge: Polity Press

Bauman, Zigmund. 2012. *Liquid Modernity Revisited.* Cambridge: Polity Press.

Berry, M. 2011. Poetic Tweets. *Text Journal of Writing and Writing Courses, Australian Association of Writing Programs, Australia* 15 (2): 1–13.

Berry, M. 2015. Out in the Open: Locating New Vernacular Practices with Smartphone Cameras. In *Studies in Australasian Cinema*, vol. 10, no. 1, pp. 53–64. ISSN: 1750-3175. UK: Taylor and Francis.

Bolter, Jay, and Grushin, Richard. 1999. *Remediation: Understanding New Media*. Cambridge, MA: The MIT Press.

Burgess, J. 2006. Hearing Ordinary Voices: Cultural Studies, Vernacular Creativity and Digital Storytelling. *Continuum: Journal of Media & Cultural Studies* 20 (2): 201–214.

Burgess, J. 2008. All Your Chocolate Rain Are Belong to Us? Viral Video, Youtube and the Dynamics of Participatory Culture. In *The VideoVortex Reader*, ed. G. Lovink and S. Niederer, 101–111. Amsterdam: Institute of Network Cultures.

Goffman, E. 1963. *Behavior in Public Places: Notes on the Social Organization of Gatherings*. New York: Free Press.

Goggin, G. 2006. *Cell Phone Culture—Mobile Technology in Everyday Life*. Oxon: Routledge.

Hjorth, Larissa, and Sarah Pink. 2014. New Visualities and the Digital Wayfarer: Reconceptualizing Camera Phone Photography and Locative Media. *Mobile Media & Communication* 2 (1): 40–57.

Horst, H.A., and D. Miller. 2012. The Digital and the Human: A Prospectus For Digital Anthropology. In *Digital Anthropology*, ed. Horst and Miller. London: Bloomsbury.

Humphreys, M., and T. Watson. 2009. Ethnographic Practices: From 'Writing Up Ethnographic Research' to 'Ethnographic Writing'. In *Organizational Ethnography: Studying the Complexity of Everyday Life*, ed. Sierk Ybema, Dvora Yanow, Harry Wels, and Frans H. Kamsteeg, London: Sage. Also available at: http://www.corwin.com/upm-data/26764_03_Ybema_Ch_02.pdf.

Ingold, T. 2007. *Lines: A Brief History*. London: Routledge.

Ingold, T. 2008. Bindings Against Boundaries: Entanglements of Life in an Open World. *Environment and Planning A* 40 (8): 1796–1810.

Ingold, T. 2010. Footprints Through the Weather-World: Walking, Breathing, Knowing. *Journal of the Royal Anthropological Institute* (N. S.): S121–S139.

Jenkins, H. 2009. *Confronting the Challenges of Participatory Culture—Media Education for the 21st Century*. Cambridge, MA: The MIT Press.

Keep, D., and M. Berry. 2013. Remediating Vertov: Man with a Movie Camera Phone. *Ubiquity: The Journal of PervasiveMedia* 2 (1): 164–179.

Lorimer, H. 2005. Cultural Geography: The Busyness of Being 'More Than Representational'. *Progress in Human Geography* 29 (1): 83–94.

Malinowski, B. 1923. The Problem of Meaning in Primitive Languages. In *The meaning of meaning*, ed. C.K. Ogden, and I.A. Richards, 146–152. London: Routledge and Kegan Paul.

Marwick, Alice, and boyd, danah. 2010. I Tweet Honestly, I Tweet Passionately: Twitter Users, Context Collapse, and the Imagined Audience, *New Media & Society* 13 (1): 114–133.

Miller, Vincent. 2008. New Media Networking and Phatic Culture. *Convergence* 14 (4): 387–400.

Okabe, D., and M. Ito. 2006. Everyday Contexts of Camera Phone Use: Steps Toward Technosocial Ethnographic Frameworks. In *Mobile communication in everyday life: An ethnographic view*, ed. J. Hofflich and M. Hartmann. Berlin: Frank and Timm.

Panagakos, Anastasia N., and Heather A. Horst. 2006. Return to Cyberia: Technology and the Social Worlds of Transnational Migrants. *Global Networks: A Journal of Transnational Affairs* 6 (2): 109–124.

Pink, S., and L. Hjorth. 2012. Emplaced Cartographies: Reconceptualising Camera Phone Practices in an Age of Locative Media. *Media International Australia* 145: 145–156.

Postill, J., and S. Pink. 2012. Social Media Ethnography: The Digital Researcher in a Messy Web. *Media International Australia* 145: 123–134.

Radovanovic, Danica, and Ragnedda, Massimo. 2012. Small Talk in the Digital Age: Making Sense of Phatic Posts, published as part of the #MSM2012 Workshop proceedings. Available online as CEUR Vol-838, at: http://ceur-ws.org/Vol-838 #MSM2012, April 16, 2012, Lyon, France. http://ceur-ws.org/Vol-838/paper_18.pdf. Accessed 10 Oct 2014.

Schleser, M., G. Wilson, and D. Keep. 2013. Small Screen and Big Screen: Mobile Film-Making in Australasia. *Ubiquity: The Journal of Pervasive Media* 2 (1–2): 118–131.

Thrift, Nigel. 2008. *Non-Representational Theory: Space, Politics, Affect*. London: Routledge.

Vannini, Phillip. 2015. 'Non-Representational Ethnography: New Ways of Animating Lifeworlds.' *Cultural Geographies* 22 (2): 317–327.

Vickers, R. 2013. Mobile Media, Participation Culture and the Digital Vernacular: 24-Hours in and the Democratization Of Documentary. *Ubiquity: The Journal of PervasiveMedia* 2 (1–2): 132–145.

Wittel, A. 2001. Toward a Network Sociality. *Theory, Culture & Society* 8 (6): 51–76.

CHAPTER 3

Performing Selfies with Smartphones

Introduction

Let me open this chapter a short vignette drawn from my ethography:

> It's Sunday afternoon. Kelly is at café with vintage décor and quaint crockery evoking a Miss Marple style era. She eavesdrops on a conversation at the next table. "Such a cute teapot, "one young woman tells her friend as they have afternoon tea. The two friends snap some photos on their smartphones including selfies with the sumptuous little cupcakes; apply faux vintage filters, kitsch illustrative overlays and share them onFacebook and Instagram with the caption 'wish you were here'. They count the 'likes' they receive in response to their posts. They laugh and discuss the jealous banter from their friends. Kelly smiled to herself thinking that people desperately want other people to be envious of them and their incredibly ordinary lives.

Kelly had recounted this commonplace event she had witnessed to me in an interview. It was in response to my question asking her to describe a typical event she'd seen where people use smartphones to take selfies in social settings. She is a thirty something photographer who uses social media both professionally and personally. The main networks she uses are Instagram, Facebook and Twitter. She describes her online persona as extraverted and she enjoys pushing boundaries. Kelly confessed that her life is nowhere near as dramatic asthe one she portrays on Facebook. I asked her about whether she takes selfies. She said that she uses selfies to

© The Author(s) 2017
M. Berry, *Creating with Mobile Media*,
DOI 10.1007/978-3-319-65316-7_3

maintain her online persona. She said that she had participated in a "take a photo of yourself for 365 days" challenge through a Twitter hash tag a couple of years ago and has continued this practice. She coyly reported that she is not sure why she continues to do this but she likes to share photos of what she is wearing as well of photos of herself feeling happy or sad. She wants people to know how she is feeling. She enjoys the feedback and acknowledgement she receives about the selfies so she keeps posting them. She feels that this practice also supports her professional life and the need to publicize her work in a subtle way. She estimated that she checks Facebook, "at least or twice once an hour" and told me, "I post to Twitter at least 15–20 times a day, more, if I am involved in an interesting conversation". She feels that many other people post selfies just to show off so that other people will feel jealous of how much fun they have every day.

But is that all there is to selfies? When I planned the content to include in this book, I felt it would be incomplete without a chapter devoted to selfies for as Gómez Cruz (2017) observed that inside the timeframe of "just a few months, the selfie, as a practice, was positioned as one of the most important and studied manifestations of digital culture" (302). A quick search in Google Scholar using the keyword "selfie" brought up an abundance of academic papers in journals. On 30 January 2017, Google Scholar returned about 9340 results in 0.07 seconds. It seems selfies have drawn much attention for scholarly research as well as providing welcome grist for popular and mainstream media columnists. One of the top entries in my search showed me that prominent new media scholar; Lev Manovich has also extended his work into selfies with a database as a key method of enquiry aptly called *Selfiecity* (selfiecity.net). The focus of his study is on the patterns that emerge through aggregating images. The project studied 3200 selfie photos from Bangkok, Berlin, Moscow, New York and São Paulo with 640 from each city. The study found that in each city studied, more women than men post selfies. Other key findings from Manovich's study were that more women struck expressive poses; people taking selfies in Bangkok and São Paulo are more like to smile that people in Moscow; and more older men post selfies than older women in each of the cities studied.

As I sat down to write this chapter, I thought about how everyone has quite a passionate opinion about selfies and a theory about why people take them and the kinds of people that do take them. The participants in my ethnography had been quite vocal when I asked for their thoughts about selfies. My observations of people in public places such as cafes and bars in various cities and villages in Europe and Australia revealed that the practice of taking group selfies was a routine part of socializing with friends. There was also an abundance of selfies on my Instagram and Facebook timelines.

In the previous chapter, I examined creative visual vernacular practices, and clearly, selfies are a subset of such practices. The possibilities of selfies are huge and almost uncontainable. Their functions and intentions include postcards, points of contact, showing off, self-affirmations (self-expression), self-promotion, self-documentation, self-invention and reinvention. The reasons why people post selfies are irreducible complex constellations and they are subject to normative pressures and local conditions. In this chapter, I look at selfies as a creative visual vernacular using a non-representational theory approach to play with mirrors: to ask when we post selfies, what do we want to see? Perhaps the answer to this depends on why we want to play with mirrors. In the next section, I address this question through a focus on some of the key functions of selfies.

Mirrors and Functions

Selfies are a flourishing area of self-expression as well as visual socialities and are increasingly coming under the gaze of academic scrutiny across several disciplines as well as a being source of discussion, debate and entertainment in popular media. Bourdieu (1990/1965) in his famous essay called *Photography: A Middle Brow Art* argued that social norms and rituals bound what was regarded as suitable or appropriate subject matter for photography in the 1960s in the era of analogue cameras and film. He observed that the occasions and composition of photographs were governed by social conventions. Formal social occasions such as weddings demanded formal lines and rows of people whereas informal events allowed for looser more creative arrangements of people. Cameras, film and processing were relatively expensive and therefore were treated with a sense of social respect. In other words, what was photographable or worth of being photographed was determined by social conventions, and to step outside of these boundaries was seen to be a waste of good film. Furthermore, photography's central function was as a souvenir to reinforce family integration as well as to celebrate leisure and social identity.

These considerations have shifted substantially in the smartphone era. Photography as an art practice has matured but with the arrival of smartphone cameras, image processing apps and social media, many banal photographs are no longer trapped in albums gathering dust somewhere on a shelf, instead they are out in the open in people's social media. Selfies have emerged as a social phenomenon out of this displacement. Selfies along with other photographs of the banal are now out in the world with their own ethos of animation, or as Gomez-Cruz and Thorman would have it, "the selfie mobilises technologies, cultural norms, and codes that are increasingly embedded in social networks, so that the practice of the selfie operates in a much wider continuum that we need to consider" (2015, 5). It is little wonder there are so many popular media articles decrying the selfie phenomena as narcissistic given that social and cultural norms have been disrupted. For example, Pearlman (2013) described the selfie phenomenon as "a symptom of social media-driven narcissism", and Hinde, a journalist with the *Huffington Post* quotes Jennifer Saunders as saying that "she believes young women are making themselves 'ill' because they spend so much time taking selfies, influenced by celebrities such as Kim Kardashian" (2016, n.p.).

The idea that the selfie phenomenon is driven by narcissism is seductive and the use of selfies by celebrities such as Kim Kardashian perpetrates this myth. Indeed, the Kardashian effect has resulted in a profusion of blogs with advice about how to take flattering selfies. In her ethnographic work on Singaporean micro-celebrities, Abidin (2014) discovered that the craft of taking selfies is treated very seriously where these women would enhance themselves cosmetically to gain "likes" and increase their base of followers by "adopting porcelain skin tones, enlarged dark pupils, and blond hair" (123). More recently her work has focused on how Singaporean micro-celebrities use selfies as a way to advertise particular products for their advertising clients. The tactics used include embedding products and logos but she observed that some of the "most labored and convincing commercial selfies occur when Influencers are photographed actually using the product or service, especially if it is in the aesthetic of a 'how to' tutorial" (2016, 8). To say that such selfies are simply narcissistic is naive. There is a clear economic imperative at work in the selfie behaviours observed by Abidin (2014, 2016) in her ethnographic research. The selfie genres she has described are reflective of the social conventions and aspirations of Singapore, and they are governed by local social norms, expectations, aspirations and conventions.

However, not all mainstream and popular media commentators see selfies as narcissistic and therefore negative. *Huffington Post* journalist, Lindsay Holmes, reports on a study conducted by researchers at the University of California (Chen et al. 2016) that found that "snapping selfies and sharing images with friends had a positive effect on their psychological and emotional states" (Holmes 2016, n.p.). The study comprised three experimental conditions (Chen et al. 2016) where their participants "were instructed to take one photo every day in one of the following three conditions: a selfie photo with a smiling expression, a photo of something that would make oneself happy and a photo of something that would make another person happy" (1). The study was premised on self-perception theory, which proposes, "how people behave will determine what they think and how they feel" (3). So if you want to feel happy, behave like a happy person. There is an element of "fake it until you make" it in this theory. If we accept self-perception theory then taking selfies with smiles can make us feel happy and indeed the results of this small-scale study conducted by Chen et al. (2016) suggest that this is indeed could be the case. Here, taking selfies has the function of focusing attention on a specific state of being or on creating an illusion of that state of being. Either way, the action of taking a selfie performing a happy face could well have the effect of nudging a person towards that state of being, in this case feeling happy.

It seems there are many theories circulating popular mainstream media as well as within academia as to why we take selfies, and how we can get a competitive edge from this phenomenon. Performing the self and self-expression is a recurrent theme that draws much interpretation and debate. Perhaps it is as Murray (2015) proposes "Ubiquitous on social media sites like Facebook, Tumblr, Flickr, and Instagram, the selfie has become a powerful means for self-expression, encouraging its makers to share the most intimate and private moments of their lives—as well as engage in a form of creative self-fashioning" (2015, 490). But in order to further explore the functions of selfies in the mirror of social media where some seem to be perpetually asking "mirror, mirror on the wall …" we need to contextualize them within mobile social media ecologies as well as within self-portrait and photographic practice genealogies.

Selfies are a part of a broader recent phenomenon of visual posting to social media. A comprehensive multi-sited study into the reasons why people post photos on social media led by Daniel Miller uncovered fifteen reasons.[1] Through this pioneering study of actual actions and practices, it was discovered that there were many genres of selfies and that these varied according to the field site of the study. Miller and Sinanen (2017) noted that photographs posted to Facebook are an integral part of people having fun and that the paradoxical relation between a transient moment and a desire to freeze the moment still drives the need to for subjects to pose albeit informally, so that photos "show people attempting to look more spontaneous, but upon inspection it becomes apparent that this 'spontaneous' look is almost as repetitious and rule-based as the prior era of formal posing" (2017, 11).

Miller's field site within the overall study was in South England and focused on teenagers. In this site, Miller found that the teenagers took three types of selfies: selfies where the person tried to look their best, groupies with friends and "uglies" where they pulled humorous faces. Selfies for public display on Instagram are carefully crafted whereas "ulgies" are deliberately humorous and are taken for private sharing on Snapchat and indicate trust within a friendship group. In my opinion, however, that as well as being markers of a trust relationship, there is also an argument that "uglies" are just as stage-managed as selfies and are governed by local norms in a similar way to selfies where people are putting forward their best face. They are still a performance of the self and are a way of managing impressions and are tied to local conditions. Miller and Sinanen argue that selfies reflect local cultural norms and that "a comparative perspective is important when considering conclusions from the trajectory of a single case" (2017, 19) so that what may be the case in South England is not the case in Trinidad or South Italy. The multi-sited study led by Miller provides ample evidence for the truth of this argument; nevertheless, there are some commonalities with regard to the reasons why people post selfies in social media as well as to the kinds of broad functions a selfie may fulfil.

The summaries on the project website about why we post to Facebook (2016) revealed that in South Italy and Trinidad where people try to look their best in fashionable clothes are a focus for selfies. The fieldwork in Brazil also found an emphasis on looking one's best but the selfies tended to be taken in the gym. In rural China girls decorated their selfies with illustrations such as hearts and stars, and in industrial centres in China, the young men depicted themselves with "hairstyles that made them look taller" reflecting their concerns with height.[2] In Chile, they discovered that people post photos of their feet in lieu of a their faces to create a "footie" which Miller et al. claim is unique to this region. The ordinariness of selfies in Chile is governed by prevailing social norms:

> In some cases the normative rules of self- presentation are not about looking your best, but simply about presenting the self in line with social expectations. This emerged very clearly in the contexts of northern Chile and the English village field sites; here it is the prevalence of a simple ordinariness that is most striking in the posting of images. Typical selfies in northern Chile, even for young people, are usually taken in their home or a friend's, at work or during a brief outing; they do not give the sense of much glamour. (Miller et al 2016, 163)

A similar visual genre exists in south Italy but with a key difference—the footie is taken when one is at leisure and in a glamorous setting such as the beach so that it "symbolises the presence of the individual within the spectacle and beauty of nature (Miller et al. 2016, 165). The genre in Italy takes on a postcard quality and is part of a creative vernacular. The selfies and footies serve an evocative intention about experiencing life out there in the world as if to say. I'm here and I'm doing stuff, having adventures and I want to share the atmosphere, ambiance, sights, smells and sounds. Pink and Hjorth (2014) put it this way in their discussion of how smartphones are used to evoke places: "Camera phone practices provide new ways of mapping place just beyond the geographic: they partake in adding social, emotional and psychological dimensions to a sense of place" (2014, 42, 45). Such images are often in settings that are out of the ordinary and may be traced back to the rituals of sending postcards from exotic or extraordinary destinations when people have left the beaten track of their everyday routines.

I return to Miller's study to explore further the reasons why people post selfies and to situate my own ethnographic research. The fieldwork in Turkey found that the practice of posting selfies to Facebook was extremely rare because they saw Facebook as a formal public space and did not feel selfies were appropriate. Indeed, according to Miller et al. (2016), in Turkey people photograph pictures of food and tables instead of each other at family gatherings thus making these events visible in a socially acceptable way. The fieldwork in both Brazil and Trinidad revealed: "categories of social class are commonly claimed visually through association with branded goods" (Miller et al. 2016, 155).

Marwick, in her study of popular Instagram users, has also noted that selfies document "what many young people dream of having and the lifestyle they dream of living" (Marwick 2015, 155). This aligns with the event related to me by Kelly in the opening vignette where she and her companion took pictures of cupcakes to show that they were at a quirky café eating highly desirable and enviable food. Their aim was to instigate a friendly envy for their lifestyle in their friends and followers.

Selfies and group selfies with expensive and desirable consumer items or those set in glamorous locations have the function of showing off. This function is readily visible on Facebook timelines and includes snaps of smiling families, bulging tables and happy lifelong buddies. These are often group selfies because it is difficult to express laughter and exuberance if one is alone. One of my participants told me the story of how she went out with a particular girlfriend to a food and wine festival. Her girlfriend kept wanting to post selfies of the two of them and would take ten to twenty shots each time and carefully choose the one where they looked their exuberant best and would add a caption saying how lucky she was to be spending a day with a childhood friend eating beautiful vegan food and drinking expensive local organic wine. She'd giggle and say how jealous her cousins would be.

My informant said she found it all a bit embarrassing because the constant selfie taking by her friend drew lots of attention from bystanders and was very disruptive to their conversations. She felt she was an object on display like a "performing seal" rather than enjoying a festival in the company close friend. People started photo-bombing them—pulling silly faces behind them. Her friend was not impressed by this but my informant found the photo-bombers most amusing and engaged in banter with them. She encouraged her friend to post the photo-bombed photos as well teasing her that they might become viral and that famous people like Prince Harry photo-bombed people.

See Me Now, Alive and Clicking

In my observations of social media timelines, I noticed that there can be a phatic or "you and me" function to selfie posts where the post is motivated by an awareness of digital co-presence and a hope that someone will engage. These posts are characterized by a sentiment of I am here, doing this—where are you and what are you doing? These selfies and group selfies are more than representational and can serve as an invitation to a conversation. As an auto-ethnographic experiment to test my notion that selfies may fulfill a "you and me" or phatic communicative function I posted a footie to my Facebook. I was at the beach on a hot afternoon. I anticipated there would be people digitally co-present on my social media timelines and that some would respond or even engage in conversation using comments. I was also thinking about selfies as a form of self-affirmation and wondered what the participants in my digital ethnography thought about this. I'd also noticed that footies had become part of the creative vernacular in my social media circles so I chose this form to invite a conversation on the topic of selfies and self-affirmation. I was at the beach on a hot summer's day (Fig. 3.1).

I received numerous likes but only one comment saying "nice toes" and this response ties in with the kinds of footies Miller et al. (2016) had uncovered in south Italy. I inferred from the responses that I'd been too oblique to draw responses about self-affirmation and phatic relations so I posted another selfie with a direct question "Are selfies about affirming being alive?" The comments came swiftly. My respondents offered theories and light-hearted banter. One who is a fellow scholar of mobile media said: "I think they have lots of reasons. Cultural capital (look what I have or where I am), I was here and this is me now. In some ways they are markers because we know that nothing is static and we are looking for some way of tracking our adventure as we age. I really was here! Lol". Another more teasing response to my selfie post and question came from a friend in the UK (Fig. 3.2).

Arguably, phatic aspects of my footie and associated question transcended the representational aspects of the visual image and the image participated in a shared social media lifeworld that was not bounded by geography or time zones. Instead, my footie functioned in a relational way to set up you and me or phatic dynamics between myself and my digitally co-present interlocutors. This gave me the impetus to pursue selfies as having self-affirmative and phatic dimensions further. I returned to my field notes and screenshots of mobile and social media timelines.

I had been observing another of the participants, Iris in my ethnography who lives on the west coast of the USA. Iris regularly posts series of images she captures on her walks. She uses pictures of shadows of herself and pictures of flowers to start a conversation with her stepdaughter Rosa who lives near London. Her stepdaughter invariably responds by providing botanical names for the flowers and then the common name. Rosa also posts images of flowers and her own shadow and seems to

At the beach

Fig. 3.1 Footies as creative vernacular for self-affirmation. Photo by Marsha Berry

always know the Latin names of the flowers and makes a point of this. They both post photos of flowers in their respective gardens, and it often turns into a gentle point scoring competition with them posting flowers to each other's posts and Rosa always wins the naming game. Villi (2016) has also conducted an ethnographic study of how people share camera phone photographs, and he also noted that many of his respondents use photographs as a form of visual phatic communication to let each other know what they are doing as well as their whereabouts.

Fig. 3.2 Conversation about selfies and footies. Screenshot by Marsha Berry

So far in this chapter, I have focused on the functions of selfies as postcards, points of contact, showing off, self-affirmations (self-expression), self-promotion, self-documentation, self-invention and reinvention. In the next section, I examine how selfies are proving to be a powerful impetus for creative practice and shift my focus from functions to artistic intentions.

Through the Glass of Artistic Intentions

In my ethnography, I found evidence that selfies have functions other than aesthetics and representation where people post selfies with creative intentions and as a conversation. My finding echo Katz and Crocker (2015) who contend,

> As such, "selfies as conversation" constitute a major step forward in visual communication within contemporary culture. This is a significant finding, since the dialectical communicative nature and entertainment value of selfies have been too often ignored in scholarly research to date. (Katz and Crocker 2015, 1862)

In addition to the functions I have discussed in the previous section—postcards, points of contact, showing off, self-affirmations (self-expression), self-promotion, self-documentation, self-invention and reinvention—entertainment emerges as another significant function for those who would use smartphones for their artistic practice.

The idea of storytelling is pertinent to selfie phenomena—what kinds of stories are people telling in their social media timelines? Which character archetypes are coming out to play? There are stories designed to inspire envy, compassion, indeed, all the human vices and virtues come out to play on social media timelines. To better understand selfie phenomena, we need to look more closely at media ecologies and communicative contexts and to situate the examination of selfies within broader social and cultural frames rather than pick them out as isolated objects of study (Miller et al. 2016). If we look deeper into the mirror, it becomes apparent that selfies have a partially shared genealogy with other self-reflective creative practices such as self-portraiture and autobiography.

A more historical perspective inflects Rettburg's theorizing of the selfie phenomenon where she places selfies on a continuum of artistic practices: "With digital cameras, smart phones and social media it is easier to create and share our self-representations. But self-representations have always been part of our culture"(2014, 2). She argues that the selfie phenomenon is part of a long history of self-expression including autobiography and self-improvement through keeping diaries and other forms of personal writing. Rettburg connects selfies with other forms of quantifying the self and self-documentation such as wearable fitness tracking devices and this is where her research becomes noteworthy because she situates selfies within a wider matrix of creative expressions and impressions being coupled with empirical traces and evidence.

Rettburg (2014) approaches selfie phenomena from a representational perspective and conceives two categories—the selfie as self-expression and self-representation—the first is communicative in function whereas the second is more about the content of the image itself. She defines selfies as visual identity markers with similar attributes to social media profile images to claim "we present a different version of ourselves in each profile picture we choose. In social media we not only present different fronts to different groups of people, as Goffman described in his foundational work on self-presentation (Goffman 1959), but we also change our self-presentation over time" (Rettburg 2014, 42). However, self-presentation and self-representation are not quite the same or equivalent. Self-presentation frequently has elements of performance (Schechner 1985; Goffman 1959) or performative (Austin) dimension whereas self-representation is about the semiotics of the image. We can gain worthwhile insights if we approach the selfie phenomenon through an ethos of animation that accounts for the backgrounds of the everyday lives that provide the impulses to perform selfies with smartphone assemblages.

Selfies are dependent on mobile media and live in a dynamic environment. Social media such as Facebook and Instagram have provided us with media ecologies where visual images take on elements of performance as well as communicative functions. They have created very fertile fields for creative practice interrogating global selfie phenomena and shifting cultural frames. We can see evidence of the fascination with selfies as an emergent and important conceptual theme of self-expression in the attention they are now receiving in international prestigious art gallery spaces. For example in March 2017, the *Selfie to Self-Expression* exhibition at the world-renowned Saatchi Gallery explored the significance of the selfie as art. The chief curator of the gallery, Nigel Hurst, sees the selfie as a reflection of how technology is used for self-expression. He has curated a show where he has placed self-portraits by Vincent Van Gogh and Tracy Emin alongside selfies by former leaders Barak Obama, David Cameron and the actor Benedict Cumberbatch. Hurst is quoted in a *Guardian* article from 23 January 2017 as saying:

> In many ways, the selfie represents the epitome of contemporary culture's transition into a highly digitalised and technologically advanced age as mobile phone technology has caught up with the camera. The exhibition will present a compelling insight into the history and creative potential of the selfie.[3]

It would seem that that change and continuity is also an important conceptual frame for how the international art community seeks to understand and engage with the selfie phenomenon. In the same article from *The Guardian*, Glory Zhang, of Huawei, said:

> The smartphone has become a tool of artistic expression. The selfie generation is becoming the self-expression generation as each of us seeks to explore and share our inner creativity through the one artistic tool to which we all have access: the smartphone.[4]

The world of the prestigious art gallery has embraced the selfie as an art form and the smartphone as a means of production. What is of significance here is that a creative vernacular whose hallmark engagement with the sociotechnical assemblages of smartphones is having such a profound effect on how photography as an artistic practice is being thought of by creative industry entities within the international art world. Mundane social practices and visual creative vernaculars are being absorbed into elite cultural institutions and this has caused a shift in how photography is being regarded. Photography is no longer thought of as just representation through the lens of semiotics; it is also thought of in terms of the embodied and social relations it generates. Frosh (2015) puts it this way:

> ... the selfie is a "gestural image" and that we should not understand its aesthetics purely in visual terms. Rather, selfies conspicuously integrate still images into a technocultural circuit of corporeal social energy that I will call kinesthetic sociability. This circuit connects the bodies of individuals, their mobility through physical and informational spaces, and the micro-bodily hand and eye movements they use to operate digital interfaces (Frosh 2015, 1608).

Each selfie is inscribed with the corporeality of the photographer and the trace of the embodied gestures of an arm raised and eyes looking into a smartphone screen to find the right angle for the shot. Saltz also notes that "selfies are nearly always taken from within an arm's length of the subject" and that there "is the near-constant visual presence of one of the photographer's arms, typically the one holding the camera (2014, 4)". Selfies are a performance and a sociable communicative practice with a first person perspective as well as being indexical and reflective. The Australian conceptual artist Jesse Willesee has adopted the selfie stick, which allows one to take selfies from a distance longer than one's arm for his project that challenges the conventions of art galleries and museums that frown on selfie sticks. He plays with the notion that curators have the means to dictate how viewers will engage with artworks in gallery spaces. He told Leigh Weingus for Huffington Post,

> After I started posing with the artworks, people wanted to pose with me. Galleries don't accept that that's the kind of experience audiences want to have. They have an old-fashioned solemn, contemplative viewing experience in mind when people want to interact and jump into the picture. Let the people take photos! (Willesee in Weingus 2015, n.p.)

This aligns well with the argument put forward by Gómez Cruz and Thornham where "we need to understand the phenomenon of the selfie as a performative and mediatory practice that cannot be reduced to, or solely taken from, the image 'itself'" and that "image-creation (along with distribution and its use in social media), does not only *represent* bodies, it also *generates* them" (Gómez Cruz and Thornham 2015). A selfie taken with an artwork invariably not only represents the artwork photographed but also creates a new artwork that includes traces of an embodied experience. The action of taking and sharing such images may also playfully disrupt the tacit norms associated with prestigious gallery spaces.

Suler (2015) approaches the practice of selfing through taking selfies through the lens of psychoanalysis and the creative practice of self-portraiture. He notes that there is a strong connection between the reactions people get when they post images of themselves to social media environments and that people will "often design a self-portrait with particular viewers in mind" and to play part such as "the loving parent, the jokester, the athlete, or the lonely sensitive introvert" (Suler 2015, 178). In other words, they post selfies in guises (and disguises) in which they would like to be seen. He argues that people who take on 365 day selfie projects, which I suggest have become a prominent creative visual vernacular in social media, can discover or rediscover what they wish to express "when they undertake ongoing self-portraits as a genuinely self-reflective process of seeing where they have been and where they might be going, the road can take unexpected twists and turns" (Suler 2015, 180). Murray (2015), in a recent article examining how young women in their teens and early twenties use selfies as a way of defining themselves, argues that "the selfie has become a powerful means for self-expression, encouraging its makers to share the most intimate and private moments of their lives—as well as engage in a form of creative self-fashioning" (Murray 2015, 490).

The unexpected twists of looking into the mirrors of selfies as a form of creative self-fashioning or a performance of the self are found in the work of Molly Soda. She is an American artist who is playing with the socially disruptive tendencies of selfies and has adopted the role of the ingénue with a project called "Should I send this?" by calling it "a collection of text and images I would be too scared to show you" (Frank 2015b, n.p.). She constructs herself as what Goldberg would call an irresponsible subject (Goldberg 2017). The collection included nude selfies and NSFW texts and sexts and was posted in the form of an electronic zine. Her project engages with emergent discourses about privacy, intimacy and vulnerability in an age of smartphone and mobile media. The series drew negative and at times aggressive comments and eventually Soda responded saying, "If none of my photos had been nudes and there had only been the text I included in my zine (which is 50% of [the zine]) no one would be calling me vapid or trash. Doesn't that have something to say about us as a society and the way we view women's bodies [and our thoughts on] them having control over their bodies and the way they choose to share it?" (Frank 2015b, n.p.). Soda uses selfies to turn the mirror back onto society to challenge norms and conventional ways of thinking about narcissism and self-imaging.

Goldberg (2017) approaches the selfie through a psychoanalytical frame to unravel why so many denounce selfies as narcissistic and suggests that "as a gesture, the selfie produces not only the enacted, post-authentic self at its centre, but a particular mode of relationality as well" (Goldberg 2017, n.p.). He draws on queer theory as well to challenge hegemonic ways "to problematize the diagnosis of narcissism as rooted in a normative project that works to produce responsible subjects, and to suggest that this project is compromised by a queer indifference to difference, as critics fear" (2017 n.p.) and concludes his exposition of contemporary preoccupations with the narcissistic dimensions of selfies by claiming, "In the selfie, critics see not only a vehicle for our superficiality but also, alongside this, a mechanism for our detachment from social bonds" (2017 n.p). A critical perspective informed by queer theory provides ways to challenge supposedly "normal" reactions to situations. The comments posted as reactions to the material in Soda's project "Should I send this?" are evidence of how passionately some feel about the maintenance of social norms governing what may be spoken and shown in publics spaces and the threat Soda's images poses to the parameters of what constitutes social acceptability.

The idea about selfies being a way to distance oneself from a situation through self-reflexivity as expounded by Goldberg where they "show a self, enacting itself" (2015, 1621) finds resonance in the Anti-Selfies series of Dutch artist Melanie Bonjano. She is a strong proponent of images that disrupt, disturb and confront dominant themes in society. In response to a question asked by Pricilla Frank about selfies and Bonjano's "Anti-Selfie" series, Bonjano said, "I started to take the pictures of myself crying as a way of putting myself outside of the situation of sadness and looking at myself from the point of view of an observer. As soon as you point the camera to yourself, you take yourself out of the moment and look at the situation from the perspective of an observer" (Frank 2015a, n.p.).

The character archetypes coming to play with selfies in art circles are numerous and conceptual artists seek to disrupt popular conceptions through performing such archetypes in their image-making practices using selfie forms. Artists such as Bonjano draw on queer and feminist theories to provide social critiques that engage with prevalent themes and discourses in the international art world. To put it another way, selfies and everyday visual vernaculars circulating in mobile media ecologies have been appropriated to provide mirrors through which artists and art institutions can see themselves.

BACK TO THE LOOKING GLASS

I began this chapter with a question: When we post selfies, what do we want see? Selfies have become grist for conceptual artists yet they have their origins in quotidian practices of sociability and are emplaced visualities (Pink and Hjorth 2012). The selfie as an object of investigation and discussion is a point of convergence between disparate disciplines. Selfies may be examined through different sets of practices and materialities, and through the lens of different agendas. For example, popular and mainstream media have framed selfies as unbridled narcissism and as being symptomatic of all that is wrong with the world. The conceptual artists that participate in the international art world have embraced selfies for their disruptive potentialities. Yet selfies by definition are networked images and participate in what Gómez Cruz and Meyer (2012) have theorized as the fifth moment in photography associated with smartphones, and comprise complex configurations of sociotechnical practices, which "include the 'I' in the act of showing or telling something" (Gómez Cruz 2017, 304).

The "I" seems obvious but in relation to the selfie phenomenon, it is always constructed and performed. It is more elusive than it first appears—especially if it is caught in the act of performing. I shall end this chapter with a story related to me by one of the participants in my ethnography whom I shall call Jasmine. Jasmine went for a walk in a deserted park behind a graveyard and came across a young woman sitting on a swing by herself laughing out loud and videoing herself with her smartphone. The young woman stopped the swing with her feet and made gestures with her thumb that indicated typing a message of some kind and posting. The young woman stood up when she had finished and walked on with a melancholy face. She seemed oblivious to Jasmine's presence in the park. Jasmine felt she'd witnessed a performance of a lonely and bored person trying to convince her social network that she was having an exciting and carefree life. She told me she was deeply affected by what she had accidentally witnessed. She'd wondered if she should have said something to the young woman, to acknowledge her in some way. It was an ethical conundrum—should she have interfered and risked embarrassing the young woman or not? She erred on the side of polite convention and pretended she hadn't noticed anything out of the ordinary. We speculated as to what the caption accompanying the video might have been and wondered how many likes the post would receive. Was the young woman simply trying to create the impression that she was having an active, fun-filled life? We imagined that the post also could be part of the young woman's creative practice as an artist or perhaps it was a simply a defiant response to the aura of graveyard.

Notes

1. http://www.ucl.ac.uk/why-we-post/discoveries/.
2. http://www.ucl.ac.uk/why-we-post/discoveries/3-there-are-many-different-genres-of-selfie_.
3. https://www.theguardian.com/artanddesign/2017/jan/23/saatchi-gallery-to-explore-selfies-as-art-f°rm-self-expression?CMP=Share_iOSApp_Other.
4. https://www.theguardian.com/artanddesign/2017/jan/23/saatchi-gallery-to-explore-selfies-as-art-form-self-expression?CMP=Share_iOSApp_Other.

References

Abidin, C. 2014. #In$Taglam: Instagram as a Repository of Taste, a Brimming Marketplace, a War of Eyeballs. In *Mobile Media Making in the Age of Smartphones*, ed. M. Berry and M. Schleser, 119–128. New York, NY: Palgrave Pivot.

Abidin, Crystal. 2016. "Aren't These Just Young, Rich Women Doing Vain Things Online?": Influencer Selfies as Subversive Frivolity. *Social Media + Society*, April-June 2016: 1–17.

Bourdieu, P. 1990 (1965). *Photography: A Middle-Brow Art*. Cambridge: Polity Press.

Chen, Yu, Gloria Mark, and Sanna Ali. 2016. Promoting Positive Affect Through Smartphone Photography. *Psychology of Well-Being* 6: 8. doi:10.1186/s13612-016-0044-4.

Frank, Priscilla. 2015a. Meet the High Priestess of the Anti-Selfie, Dutch Artist Melanie Bonajo (NSFW). *Huffington Post.* http://www.huffingtonpost.com. au/entry/melanie-bonajo_n_5811496.

Frank, Priscilla. 2015b. Feminist Artist Leaks Her Own Nudes, Internet Responds like Meatheads (NSFW). *Huffington Post.* http://www.huffingtonpost. com.au/entry/molly-soda-nude-photos_n_7565838.

Frosh, Paul. 2015. The Gestural Image: The Selfie, Photography Theory, and Kinesthetic Sociability. *Israel International Journal of Communication* 9, Feature 1607–1628. http://ijoc.org/index.php/ijoc/article/viewFile/3146/1388.

Goffman, Erving. 1959. *The Presentation of the Self in Everyday Life.* Garden City, New York: Doubleday.

Goldberg, Greg. 2017. Through the Looking Glass: The Queer Narcissism of Selfies. *Social Media and Society* 3 (1). March 27, 2017. http://journals. sagepub.com/doi/full/10.1177/2056305117698494.

Gómez Cruz, Edgar, and Eric Meyer. 2012. Creation and Control in the Photographic Process: Iphones and the Emrging Fifth Moment of Photography. *Photographies* 5 (2): 203–211.

Gómez Cruz, Edgar, and Helen Thornham. 2015. Selfies Beyond Self-Representation: The (Theoretical) F(r)ictions of a Practice. *Journal of Aesthetics and Culture* 7 (1): 1–10.

Gómez Cruz, Edgar. 2017. The (Be)coming of Selfies: Revisiting an Onlife Ethnography On Digital Photography Practices. In *The Routledge Companion to Digital Ethnography,* ed. L. Hjorth et al. New York: Routledge.

Hinde, Natasha. 2016. Jennifer Saunders: Girls' Quest for the Perfect Selfie Is Making Them 'Ill'. *Huffington Post 17/09/2016.* http://www.huffingtonpost. com.au/2016/11/28/jennifer-saunders-believes-girls-quest-for-the-perfect-selfie-is-making-them-ill/?utm_hp_ref=au-selfies.

Holmes, Lindsay. 2016. Science Says Selfies Can Make You Happier and More Confident. *Huffington Post 17/09/2016.* http://www.huffingtonpost.com. au/2016/09/18/science-says-selfies-can-make-you-happier-and-more-confident/?utm_hp_ref=au-selfies.

Katz, James, E and Crocker, Elizabeth, T. 2015. Selfies and Photo Messaging as Visual Conversation: Reports from the United States, United Kingdom and China, *International Journal of Communication* 9(2015), Feature 1861–1872.

Manovich, Selfiecity http://manovich.net/content/04-projects/085-selfiecity-exploring/selfiecity_chapter.pdf.

Marwick, A.E. 2015. Instafame: Luxury Selfies in the Attention Economy. *Public Culture* 27: 137–160.

Miller, Daniel and Sinanen, Jolyanna. 2017. *Visualising Facebook*. UCL Press. http://www.ucl.ac.uk/ucl-press/browse-books/visualising-facebook. Accessed 24 Aug 2017.

Miller, Daniel, Elisabetta Costa, Nell Haynes, Tom McDonald, Razvan Nicolescu, Jolynna Sinanan, et al. 2016. *How the World Changed Social Media*. UCL Press. Pdf available www.ucl.ac.uk/ucl-press.

Murray, Derek Conrad. 2015. Notes to Self: The Visual Culture of Selfies in the Age of Social Media. *Journal Consumption Markets & Culture* 10 (6): 490–516.

Pearlman, J. 2013. Australian Man "Invented the Selfie After Drunken Night Out". *The Telegraph*, Nov 19. www.telegraph.co.uk/news/worldnews/australiaandthepacific/australia/10459115/Australian-man-invented-the-selfie-after-drunken-night-out.html.

Pink, S., and L. Hjorth. 2012. Emplaced Cartographies: Reconceptualising Camera Phone Practices in an Age of Locative Media. *Media International Australia* 145: 145–156.

Pink, Sarah., and Hjorth, Larissa, 2014. New visualities and the digital wayfarer: Reconceptualizing camera phone photography and locative media, *Mobile Media & Communication* 2 (1): 40–57.

Rettburg, Jill Walker. 2014. *Seeing Ourselves Through Technology: How We Use Selfies, Blogs and Wearable Devices to See and Shape Ourselves*. New York: Palgrave Macmillan.

Saltz, J. 2014. Art At Arm's Length: A History of the Selfie. Retrieved from http://www.vulture.com/2014/01/history-of-the-selfie.html.

Schechner, Richard. 1985. *Between Theater and Anthropology*. Philadelphia: University of Pennsylvania Press.

Suler, John. 2015. From Self-Portraits to Selfies. *International Journal of Applied Psychoanalytic Studies*, 12: 175–180. doi: 10.1002/aps.1448.

Villi, Mikko. 2016. Photographs of Place in Phonespace: Camera Phones as a Location-Aware Mobile Technology. In *Digital Photography and Everyday Life: Empirical Studies on Material Visual Practices*, ed. Asko Lehmuskallio and Edgar Gómez Cruz, 107–121. London: Routledge.

Weingus, Leigh. 2015. Artist Challenges Selfie Stick Ban by Taking a Bunch of Selfies. *Huffington Post*. http://www.huffingtonpost.com.au/entry/selfie-stick-ban-artist_n_6768980.

CHAPTER 4

'Being There' with Smartphone Apps

INTRODUCTION

In the previous two chapters, I discussed the emergence and formation of creative visual vernaculars. The extreme accessibility of smartphones has provided us with a hitherto unprecedented opportunity to convince our interlocutors that we have "been there" through photographic and video evidence. So, how do we imagine and reimagine places and events through the use of smartphone assemblages to evoke a sense of what it's like to "be there"? And what does it mean to "be there"? And what part does digital co-presence play in making mobile art?

Messing about with smartphone apps can result in evocative expressions that are part of the constellation of phenomena that Sheller (2014) and Hjorth (2016) have defined as mobile art. People make and share mobile art as a part of their everyday routines using the extreme accessibility of smartphone affordances and mobile phone networks. The vignette below is a narrative inference drawn from my ethnographic research:

> In the street, a man sidesteps a woman walking and, completely immersed in the contents of her smartphone screen. He, on the other hand, is keenly aware of the cool breeze on his cheeks, the smell of rain and is delighted by the sight of the shadows and reflections on the slick pavement He shakes his head smiling at her lack of peripheral vision, stops, takes a fifteen second video of her back receding down the street, applies a high

© The Author(s) 2017
M. Berry, *Creating with Mobile Media*,
DOI 10.1007/978-3-319-65316-7_4

contrast black and white filter. He sends it to Instagram and Twitter with an ironic film noir style haiku in the caption box with the hash tags #haiku, #distracted, #needstosmellcoffee, #filmnoir. Within minutes it is liked on Instagram and retweeted on Twitter by numerous people. He hasn't lost his touch.

In this vignette, we see the collision of mobility, digital co-presence and digital wayfaring. These function as preconditions for mobile art in this vignette. My respondent had told me that he enjoys creating video art and photographic works that evoke a sense of what it was like to be there in the street with distracted pedestrians. He seeks to create images with emotional resonance for people he thought might be digitally co-present depending on the way time zones would line up. He feels that camera apps have provided him with a whole darkroom embedded his smartphone and that he could play on nostalgia and faux-vintage aesthetics as ways to add emotional and atmospheric depth to things that caught his eye. In previous work, I argued that the fascination with faux-vintage apps is an indication that people are trying to communicate more than what is simply in front of their eyes and that they are trying to convey a sensory impression that will capture the imaginations of digitally co-present others (Berry 2014). Here, I build on this work to explore further how the extreme accessibility of smartphone apps and digital co-presence are seeding emergent forms of mobile art and screen production. In the next section, I discuss the relation between evocative moments and everyday life to set the scene for the rest of the chapter.

THE EVOCATIVE AND THE EVERYDAY

Photography and video have become intertwined with the embodied mobility of everyday life, and this, in turn, influences contemporary everyday aesthetics and extends art practices into new areas. The affordances offered by smartphones have given rise to a huge array of remediation (Bolter and Grusin 1999) of aesthetic expression. But there is more to mobile art than remediation and the hybridization of existing forms. Sheller (2014) points out that mobile art should be seen as much more than remediating analogue forms and much more than simply adapting art and film meant for other screen contexts to smartphone screens. She states

Mobile art has in fact expanded the spatial and social field in which art takes place by experimenting with the mobile interface as a bridge between digital and physical space, a hybrid mediation of human sensory perception and technological connectivity. (Sheller 2014, 376)

The extreme accessibility of smartphones has created fertile ground for experimentation with new forms of creative expression and actions where boundaries between digital and physical space are blurred, shifted and transfigured and where social relations may become woven into the mobile art itself. Nevertheless, the relations between social practices and mobile art remain "undertheorized and discussed" according to Hjorth (2016, 169), and in response to this apparent lack, she raises questions aimed at establishing what constitutes mobile art and whether can be defined through its use of mobile interfaces in the process of making the work or its delivery and whether its relationship to mobile content and context are defining characteristics. She argues that mobile art needs to be understood as "a broader field of creative practice than just locative media or media arts practice" (Hjorth 2016, 170) and that mobile art can "help us reconceptualise the relationship between art and creative practice in new ways" (170). She does this by using three thematic rubrics to define and theorize mobile art, whereby the "rubrics seek to consolidate mobile art practices as deploying an overlay between the networked (intimate copresence), haptic visualities like camera phone apps (emplaced visuality), and playful interventions in the everyday (ambient play)" (Hjorth 2016, 170).

This discussion has much relevance at this particular point in time because we are now privy to many aspects of everyday life that were previously difficult to access because we live in an age where "social media render the back and forth of social life perceptible to analysis" (Crang 2015, 345). The back and forth of social life includes visual images and creative expressions which are exchanged and discussed on social media timelines on an everyday basis yet rarely is this quotidian fluidity problematized and analysed. Hjorth's rubrics provide a useful starting point to explore intersections between the evocative and the everyday as they are performed in mobile media lifeworlds. Although I have discussed what constitutes a lifeworld earlier (Chap. 1), for my purposes in this chapter, I provide the following definition:

the world as immediately or directly experienced in the subjectivity of everyday life, as sharply distinguished from the objective "worlds" of the sciences, which employ the methods of the mathematical sciences of nature;
although these sciences originate in the life-world, they are not those of
everyday life. http://www.britannica.com/topic/life-world

A lifeworld, then, is relational and intersubjective. Our lifeworld
includes our complex, and at times, second nature use of smartphones
and networked technology. Our ability to easily document our movements through everyday life has shifted how we think about film and
photography.

THE VITALITY OF "BEING THERE"

Being there immersed in a specific place remains a fundamental concept
in ethnography since Geertz wrote the influential work entitled, The
Interpretation of Cultures: Selected Essays in 1973 and provides a touchstone for my discussion of mobile art practices in this chapter because it
presupposes some kind of experiential and conscious presence whether it
be physical or mediated through mobile technology. In his famous essay,
"Being There", Geertz (1988) argues that

> The ability of anthropologists to get us to take what they seriously has less
> to do with either a factual look or a conceptual elegance than it has with
> their capacity to convince us that what they say is a result of their having
> actually penetrated (or, if you prefer, having been penetrated by) another
> form of life, of having, one way or another, truly having "been there". And
> that, persuading us that this offstage miracle has occurred, is where writing
> comes in. (Geertz 1988, 5)

As I noted in Chap. 1, ethnographic fieldwork may be likened to a
hero setting off on a journey into the unknown and returning with the
elixir. Being there also implies an arrival somewhere at a point in time
and an awareness of one's presence in that somewhere. Once mobile
phones became ubiquitous in the late twentieth and early twenty-first
centuries, the idea of presence—and its corollary co-presence—became
more complex because a virtual or networked dimension became a
tangible addition. The idea of co-presence digitally and otherwise has
dimensions of vitality or liveliness and participates in an ethos of animation

(Vannini 2015) because it is always dynamic. The ease of being able to document and share our movements through everyday life has recalibrated how we think about film and photography. In this chapter, I reflect on the actions within my own recent creative practice with smartphones where I explore the implications of wayfaring, digital co-presence and mobility through video and photography. I draw on my small-scale digital ethnography as well, which I describe in some detail in Chap. 2, which was conducted between 2013 and 2016. To reiterate briefly, I observed and interviewed sixteen self-identified artists and poets about how they used smartphones and social media for their creative practice. In Chap. 2, I focused on how they made and engaged with photographs and videos on a daily basis in mobile media spaces where new vernaculars are emerging, and I pick up some of those threads here to trace how wayfaring, digital co-presence and mobility play out to create moments of emotional resonance.

The idea of co-presence was first conceptualized by Goffman (1963), whereby "copresence renders persons uniquely accessible, available, and subject to one another" (Goffman 1963, 22). It is achieved when people "sense that they are close enough to be perceived in whatever they are doing, including their experiencing of others, and close enough to be perceived in this sensing of being perceived" (17). Goffman's concept presupposed physical proximity. The notion of co-presence has changed since the arrival of mobile communications technology (see, for example, Hjorth 2016) to include networked proximity.

Let me diverge now from discussing the history of the idea of co-presence for a moment to present a brief account of work travel to Vietnam with my smartphone in 2010 to unpack some of the practicalities and the vitality of digital co-presence. At the time, I was an avid user of Twitter and was investigating its potential for creative practice. I had built up a following, and there were poets and artists amongst my followers. My work colleague in Vietnam was also a prolific user of Twitter. She had a large following of over 3000 at the time of my trip. At that time, my followers numbered at round 1500. I began documenting my trip checking off various items using Twitter as I packed my case. I took a photo of my cat curled up on top of my things in the suitcase and posted it to Twitter. I remained in touch with my followers and was receiving responses right up until I had to switch my phone off for the flight.

My colleague in Vietnam had been following my progress as well and would announce at regular time intervals that I was coming over to work with her and retweet or share my posts. I continued to post photos along with the updates and began to understand that this was taking my creative practice with photography and video into new fields for me. I had already experimented with mobile phone photography and film-making with my old Nokia, but this was something else. I had direct access to social media and to my blog through Internet connectivity, so I could shoot, edit and upload images with my smartphone. The experience was giving me a sense of immediacy and a strong and lively connection with actual interlocutors rather than an imaginary model reader or viewer (Eco 1979). My smartphone camera also could do something my humble Nokia could not—with it, I could do post-production in camera using filters without having to download images to my laptop, open photo-editing software and edit the images and then post them straight to social media.

Gòmez Cruz and Meyer (2012, 216) have identified the capability of smartphones to provide access to photo-editing apps as a fifth moment in photography because we no longer need sophisticated skills in professional photo-editing software to alter the look and feel of a photograph. I was experiencing this fifth moment first-hand in 2010 and began to explore how smartphone assemblages were giving rise to new forms of photography, film-making and writing through my own creative practice as well as through digital ethnographic fieldwork. The fifth moment in photography was magnifying my sense of digital co-presence and my ability to initiate direct exchanges with my followers through creative practice—theirs as well as mine. This gave me a sense of vitality and liveliness that was new. My smartphone also provided me with the technical means to create images that would hopefully evoke emotional resonance.

I was expanding a practice that I had built previously with my Nokia which was a marriage of sensory visual ethnography using video as developed by Pink (2009) with my own video art and photography experimentation using the production value constraints of the Nokia mobile phone camera to explore low-resolution aesthetics. The smartphone—an iPhone 3—allowed me to experiment with filters to create specific effects including double exposure in camera. These effects were immediately accessible through a tactile touch screen, and the interfaces were easy to learn and felt intuitive in a remarkably short period of time. If I was in a place that had Wi-fi or mobile phone network connection, I was able to share images of street life in Ho Chi Minh City using applications like Twitter and Instagram with my followers almost straight away using my smartphone. My colleague and I collaborated on a real-time project where she mapped out a route where we took photos and posted these from cafes with free Wi-fi. Our followers responded enthusiastically with comments and retweets. My colleague remarked that she felt the audience of our followers was peeking over our shoulders as we walked, photographed, applied filters and posted to social media. For her, they were an almost psychic presence, sharing the wonders and pointing out the small details of events. Our audience of followers was experiencing our guided tour through digital co-presence.

Mobile phones and smartphones have changed our lifeworlds and how we behave in public places immutably. While smartphone assemblages and their extreme accessibility extend spaces for new and hybrid forms of creative practice, digital co-presence is double-edged. Mainstream media bemoans the time we spend with our mobile devices claiming we are increasingly disconnected from physical settings. An important contributor to the discussion of the shifting nature of co-presence and our use of mobile devices is Sherri Turkle (2008), a prominent US psychologist. She claims that our rapid cycling through mobile media creates "a sense of continual co-presence" (122) and that this creates a perpetual distraction and may cause anxiety. She stresses the adverse effects of social media and mobile devices to argue that these diminish the quality of human interactions. My concern in this chapter is not so much with the effects, adverse or otherwise, mobile communications technologies have had on communication and social relationships; rather, my interest is in how these technologies are being harnessed for creative practice through the changes in how co-presence with synchronous and asynchronous others may be experienced. Nevertheless, it behoves me to acknowledge that smartphones and social media may propel us into an in-between elsewhere where we are disconnected from our physical surroundings; however, I would argue that there are other many things as well that can distract from a sense of being present in a place or situation.

Digital co-presence is a fact of contemporary life for many, and the lines between online and offline are indistinct. Hjorth (2016) argues that co-presence rather be thought of as a spectrum or range of interpersonal engagement that "goes beyond counterproductive dichotomous models of online and offline, here and there, virtual and actual" (Hjorth 2016, 175). It is more useful to conceptualize co-presence in the form of a rubric and add in other dimensions such as synchronicity, time zones and geographic distance rather than simply in terms of physical offline presence and networked presence. When we look at social media timelines, we can readily observe emplaced visualities in operation. It is here that the unproductiveness of thinking about the space generated by social media in binaries, such as online and offline, physical and virtual comes to the fore.

Similarly, conflating conceptualizations of "being there" in the field with actual physical locations is no longer fruitful because of the changing nature of what it means to be co-present with other humans. Postill (2015) unpacked the conundrum of how to conceive of being there in a digital era in a blog post where he proposed four ways of being there—physically, remotely, virtually and imaginatively. He was unpacking the implications of what being there meant for digital ethnographic fieldwork. His four ways have significant implications for creative practice using smartphone assemblages. If we apply these four modes to thinking about co-presence, we can explore new ways of creating mobile art that speaks from an illocutionary position that truly has been there. Another useful inference to be drawn from Postill's four modes of being there in the field is that acts of the imagination "before and/or after the fact, through digital stories or images found on blogs, social media, video-sharing sites" (2015, n.p.) influence the way we experience digital co-presence, and this in turn, in my opinion, has huge implications for creative practices and evocative moments with smartphones. Postill's mode concerning acts of the imagination as a way of being there also speaks to Vannini's (2015) call for an ethos of animation in ethnographic writing and emphasizes the vitality of co-presence and being there.

LIVING TOGETHER

Co-presence and being there have philosophical implications as well because these ideas are predicated upon the concept of a lifeworld where subjectivities are shared. Let's dwell for a moment on what our oft taken-for-granted shared subjectivities or intersubjectivities of what makes up the world may imply. The phenomenologist Edmund Hursserl in his volume titled *The Crisis of the European Sciences* first introduced the idea of a shared lifeworld where we all live together in a world that is external to ourselves that we all live in and make sense of together:

In whatever way we may be conscious of the world as universal horizon, as coherent universe of existing objects, we, each "I-the-man" and all of us together, belong to the world as living with one another in the world; and the world is our world, valid for our consciousness as existing precisely through this 'living together.' We, as living in wakeful world-consciousness, are constantly active on the basis of our passive having of the world... Obviously this is true not only for me, the individual ego; rather we, in living together, have the world pre-given in this together, belong, the world as world for all, pre-given with this ontic meaning... The we-subjectivity... [is] constantly functioning. (1936: 108–109)

Our "living together" in the world has been transformed through the affordances of smartphones even though smartphones themselves can be seen as an interim technology that is "an amalgamation of familiar media along with a few new ones that are constantly being improved" (Miller 2014, 211) as well as a source of distraction that can have adverse effects on interpersonal relations (Turkle 2008). Smartphones have become a part of the world as a "universal horizon" as "existing objects" in a "coherent universe". Despite the many distractions posed by smartphones, people in many countries around the world communicate multi-sensory and heightened aspects of everyday banalities through smartphone apps that sit in the background of everyday activities and practices. The urge to feel connected to other humans remains strong, and we continue to value descriptive accounts of what it's like to be somewhere else. And, we can still feel creative impulses to share what it's like to be where we happen to be so that others may feel for us in our lifeworld.

We can find much evidence of these creative impulses in social media. For example, mobile media is awash with pictures of the sky, and there is even a hashtag—#skyporn—to bring together people who still gaze at the sky and wonder. In June 2015, Huffington Post's Suzy Strutner wrote an article dedicated to this hashtag saying that "#Skyporn is an innocent, entirely wondrous trend". Our fragile blue planet continues to inspire the poetic impulse, even in a world where attention is sucked into screens. The images posted to the popular Instagram social media app using the #skyporn hashtag are often filtered to accentuate the atmosphere which tends to be a signature aesthetic of images taken with this application. On 3 May 2016, there were 11,470,281 posts on Instagram at 1:47 pm AEST using the #skyporn hashtag, and by 17 March 2017, there were 15,996,266 posts. Kevin Systrom, a founder of Instagram, describes how he came upon the idea of ready-made filters in an interview with Hannah Kuchler of FT Magazine:

> Studying in Florence during college, he took a photography class in which a teacher pushed him to try a plastic camera and add chemicals to the developing solutions to achieve interesting effects. "That changed my life. I mean, you know, the discovery of square-format, filtered photos. I mean, that's it, right?" he says, gesturing at the plus-sized Instagram photos on the wall. (Kuchler 2015[1])

Systrom's comments about square-format toy cameras and their association with mid-twentieth-century art schools provide Instagram images with a genealogy that includes lomography and instant photography with iconic artists including Andy Warhol, David Hockney, Maripol, Robert Mappelthorp and Helmut Newton. Caoduro suggests that "Instagram emerges from a culture where the old, the authentic, the analogue is still a repository of value and appreciation" (Caoduro 2014, 73), and perhaps, it is true that the aesthetic choices we choose are a manifestation of the enduring influence of analogue values. In turn, these analogue values have become a shared frame for what Bourriaud (2002) theorizes as relational aesthetics where notions of beauty and the poetic are intersubjective. But what was once the domain of the avant-garde has now generated visual vernaculars to create moments of emplaced visuality (Pink and Hjorth 2012). Josh Constine (2013, n.p.), a technology journalist, described his first experience of using Instagram as transformational:

> But my most vivid "eureka" moment with social media happened while I was walking in Golden Gate Park at sunset. Before me, creamy cloudflare reflected off a pond. It was so beautiful it felt selfish to keep it to myself. I wanted all my friends to see what I saw.
>
> But I was no photographer, and held just a crummy early-generation iPhone. The shot lacked the vibrance and emotion of being there. Yet with Instagram's filters I could return the essence of the moment to what I captured with my camera. And with time, a community grew around the ability to be transported.

The ability to apply a filter without needing to resort to complicated photo-editing software has opened a whole new world filled with creative potential where emplaced visualities are enacted and re-enacted. Filters are a quick and simple way to evoke a desired atmosphere or ambience. Kris Fallon (2014) notes: "Filtered images do not claim 'this is how it looked' but rather 'how I wanted it to look' or 'how I felt it looked'". The uses of various filters are also popular in the worlds of fashion photography as pointed out by Picarelli. She argues that

> Retro-looking photography expresses the awareness that, in the age of social media, subjectivation is tied to the curatorial effort of producing, managing, and assembling traces of our lives for collective consumption. As an externalization of our inner life of 'moods' and 'desires,' it mobilizes subjectivation-in-the-making as a resource. (2015, n.p.)

I have argued in past work that filters associated with smartphone apps participate in a temporal loop because "smartphones have irrevocably altered and unsettled our sense of place as well as time" (Berry 2014, 63). There is a paradoxical relation between a desire for continuity and a desire to evoke the essence of the present moment and share this with the digitally co-present. I used the frame of hauntology, drawn from Derrida (1993) to support my position that the popularity of faux-vintage filters is much more than nostalgia for an idealized mid-twentieth-century past, and I proposed that we "are haunted by a lost future where we can be in a place without the unsettling presence of co-present others just about everywhere we go" and that "the popularity of faux-vintage apps is both symptomatic of our times and a product of hauntology" (Berry 2014, 63). Co-presence itself may be experienced as something spectral and unsettling as in the account of the guided tour my colleague and I conducted on Ho Chi Minh City in late 2010 where my colleague remarked she felt as if our Twitter followers were walking along with us, just behind our shoulders.

AURATIC AFFECTS

Taking photos and shooting video clips are activities that have become enmeshed with our mobility, digital co-presence and use of smartphones in our ordinary lifeworlds. The digital and spectral co-presence of others can create an auratic affect as we go about our daily business. Company is often only a click away through smartphone affordances and assemblages. The idea of mobility implies times of stillness (Bissell and Fuller 2011), being in transit (Berry and Hamilton 2010) as well as the infrastructure moorings that underpin the use of mobile devices (Hannan, Sheller and Urry 2006). Our everyday movements along the paths between both our digital and physical destinations can provide unsolicited stimuli to create and share evocative moments. I present another a vignette that I initially wrote for a conference presentation (Berry 2016) based observations of one of the participants in my digital ethnography to reveal how the creation of evocative moments over a period of time unfolded and how mobility and digital co-presence combined to produce a sense of what it means to "be there". Here, I present it to show how the digital and spectral co-presence of others combine into an everyday ritual of walking along a towpath that includes actions that make up a creative practice that produces an auratic affect:

> John, walked along a towpath beside a canal in the south of England each day on his way to the bus stop. This was a precious part of his day. On these walks, he photographed a pair of swans raising signets and posted these to his Twitter feed each day. His followers would retweet the photos and comment on the weather conditions and growth of the signets. He would reply, often in the form of haiku poems. The Twitter conversations would often turn to the subject of love and raising families. One day, he posted a photo of the canal without the swans. The photo was monochrome. He followed this up with a short video pan of the canal and its banks. Still, no sign of the swans. This triggered an outpouring of concern: was he okay? Was his family okay? Were the swans all right? Had something bad happened to the swans?

> John replied that he couldn't see the swans and was worried that foxes had taken them or that vandals had killed them. The responses were swift. People were ready to hunt down the vandals and wanted stocks and pillories brought back so that the vandals would be publically punished and humiliated. The next day he posted a photograph with the little swan family intact. The sighs of relief amongst John's Twitter followers around the world were palpable. Many smiley faces prefaced the retweets of the photo that day.

The sense of "being there" is almost palpable for John's followers. John's photographs of the swans each day had built up a shared personal context between John and his followers, which could be described as auratic in affect as well as filled with emotional resonance. Kelly has suggested that Benjamin's notion of the aura could be applied as an analogy "to the act of sifting through one's news feed in Instagram" (Kelly 2014, 132) and referred to Benjamin's oft-cited definition to support his point. I reproduce Benjamin's definition of aura here:

> A particular web of space and time: the unique manifestation of a distance, however near it may appear to be. To follow, while reclining on a summer's noon, the outline of a mountain range on the horizon or a branch, which casts its shadow on the observer until the moment or the hour partakes of their presence – this is to breathe the aura of these mountains, of this branch. (Benjamin 1931, 20)

The aura of the images posted by John each day of the swans swimming in the canal stretched a web of time and space providing his followers with a moment in their day where they could breathe the aura of the canal and its swans. I would also suggest that the popularity of the #skyporn hashtag is because the images there are an invitation to breathe the aura of a breath-taking sky; in other words, these images have an auratic affect on the viewer.

Kelly (2014) describes the place of Instagram in his own creative practice—the making of his interactive documentary *North* that charts his encounters with places in inner Melbourne—with reference to its ability to manifest an aura:

> My own Instagram account (@patcheskelly), for instance, demonstrates the ability for mobile platforms to generate a feeling of aura, particularly in a series of videos I shot for the production of an interactive documentary titled North. … The portability and high-functionality of both the device and platform mean that images can be captured promptly and inconspicuously with minimal effort, and therefore offer an even more personal experience for the capturer. As such, it was not difficult to capture moments and spaces that I found significant with ease and in a short amount of time. (Kelly 2014, 135)

For Kelly, the ease of capture and ready access to production and distribution tools in the form of mobile platforms such as Instagram contributes to the ability to generate a feeling of aura.

WAYFARING MINDFULLY

Aura and emotional resonance seem to intertwine so that places that manifest aura also have a clear emotional resonance. The ease of capturing moments and the extreme accessibility of production tools also motivated my own creative practice. Earlier in this chapter, I suggested that wayfaring, co-presence and mobility are concepts through which film-making can be reimagined. Our ability to easily document our movements through everyday life has shifted how we think about film and photography. I researched how these concepts may be used to reimagine film-making through my creative practice. I return to an important concept drawn from digital ethnography to address this question. Hjorth and Pink (2014) coined the term digital wayfarer to problematize the entanglements of online and offline worlds:

> … the digital wayfarer as we conceptualize her or him does not simply weave her or his way around the material physical world. Rather, their trajectory entangles online and offline as they move through the weather and the air, with the ground underfoot and surrounded by people and things, while also traversing digital maps, social networking sites, and other online elements. (2014, 45–46)

Opportunities for image making (Schleser et al. 2013; Berry 2013) are embedded in the background in our lifeworlds because of smartphones and camera applications. Most of us are digital wayfarers now: This has changed the game for film-makers and photographers. Digital wayfaring has paved the way for new forms and processes for film-making and photography that reflect our shared lifeworlds as we move through digital and physical spaces that are almost seamlessly woven together most of the time. With this new insight, I realized that my 2010 collaboration with a colleague in Ho Chi Minh City was a case of wayfaring both digitally and physically. My experiences using my smartphone for ad hoc creative projects in this experimental way became seeds for more practice-based research into the interstices between our lifeworlds, social media and smartphone cameras.

In the 2016 Sightlines call for papers and films, there was a question that caught my attention: "What new forms of screen production are emerging and in what ways is creative practice research engaging with them?" I felt that the extreme accessibility of smartphones and camera apps had generated new opportunities and contexts for photographers, film-makers, poets and screenwriters. I was also participating in numerous haiku groups in social media and was noticing that an increasing visual dimension was creeping in. More and more poets were adding photographs or short videos to their haiku. I observed this in the practices of the participants in my ethnography as well.

I created a short video artwork called *Wayfarer's Trail*, which was made as an experiment in the form of a walking meditation engaging with the extreme accessibility of smartphone cameras. My inspiration came from Zen philosophy and non-representational theory (Ingold) rather than from psychogeography (Debord) or the modernist notion of a flâneur as expounded by Walter Benjamin. The idea of a flâneur underpins Benjamin's monumental *Arcades Project* which seeks to provide an alternate history of modernity. Benjamin's flâneur is closely connected with "social life in nineteenth century France" (Birkerts 1982, 175). The city is envisaged as labyrinthine and layered so that the flâneur renders meanings from connections and juxtapositions using "his highly intuitive discrimination" in order to ascertain which "detail 'speaks' and he must have some sense about how it may all relate" which in turn suggests a knowledge or "some prior intuition of form or pattern" according to Birkerts (1982, 171). In other words, the purposes of flâneurs dérives are to read the city through juxtapositions and layered patterns. The figure of the flâneur as one who observes the streams of life in urban settings, moving alongside it but detached, has a desire to expose patterns and potential connections by following a set of self-imposed rules. My project was not about uncovering meanings through discerning patterns and connections.

To show how my project diverged from psychogeography, I will first define this term through an example. Nick Gadd, a Melbourne writer, has undertaken a walking project called the *Melbourne Circle,* which is an experiment in psychogeography. Gadd established some clear rules for himself:

... don't bother about famous places. Avoid the obvious. Points are earned for the unnoticed, the disregarded, the neglected. Derelict factories are good, and other places of abandon; graffiti, faded signage, street art (but not the well-known, over-hyped tourist stuff); liminal spaces, vague terrain and edgelands; sites of environmental damage or restoration. If you must look at something familiar, find a new angle. See the city as it has never been seen before. (Gadd 2015, n.p.)

My project had no such constraints or rules commonly associated with psychogeography. I worked intuitively, intentionally without any specific intention other than to simply notice my embodied and emplaced experiences while walking along a coastal trail. My only constraint was to be present with all my senses and to stop, look and capture whatever caught my eye with my smartphone. I practised a contemporary adaptation of Kinhin where it is more of an open self-monitoring observation that focuses on becoming grounded in the present moment; it is a practice that the Vietnamese Zen monk Thich Nhat Hahn describes with deceptive simplicity where "When you walk, arrive with every step. That is walking meditation. There's nothing else to it" (Hahn 2015, 12). Kabat Zinh put it another a way, "paying attention in a particular way: on purpose, in the present moment, and nonjudgmentally" (Kabat-Zinn 2005, 4). I sought to achieve a stillness of mind while walking that Minor White described as

> The state of mind of the photographer while creating is blank. ...a special kind of blank. It is a very active state of mind really, a very receptive state of mind, ready at an inst ant to grasp an image, yet with no image, preformed pattern or preconceived idea of how anything ought to look is essential to this blank condition. (Minor White, "A Living Remembrance", 1984, 36)

White wrote these words prior to the advent of digital co-presence. The psychologist Duma also commented on the absence of the thought of audience in her own photographic practice where this became for her, a marker of mindfulness:

> I began wandering with my camera in my early 20's. Often aimless in my ventures, I captured only those images that called out to me. At times, it was the strength of lines or the familiarity found in a repetitive pattern that drew me in. Other times, it was the way the light brushed against form. With no audience in mind, I sought only direct contact with the image. I found myself in a space where time lost all meaning, where the ordinary became wondrous. In the spirit of play, I saw the overlapping arches of a familiar building unfold before me, while the warmth of their colours invited me in. I was seeing it for the first time. (Duma 2012, 16).

My intention was not to see for the first time. I was not searching for an epiphany. I was interested in how the combination of smartphone apps and digital co-presence was shaping my creative practice. I located what Henri Cartier-Bresson (1990) termed "decisive moments" and observed what happened when I arrived in the moment. As I stopped to photograph or video, paradoxically yet intuitively, I imagined motion and stills, layers and double exposures and posted to social media—Twitter and Facebook—and checked how people reacted to my posts. My photos, videos and poems were pulled into a narrative about the universalities of beach trails even though the specifics were unknowable to my interlocutors. My walks became walking meditations. The stops to take photographs or jot down lines for poetry became ways of clearing the mind back to stillness. At the same time, the pauses were also a series of arrivals.

Non-representational theory provided me with a frame within which to situate my practice. In "Footprints through the weather-world: walking, breathing knowing" (2010), Ingold explored the relations between walking, the weather and knowing or being knowledgeable. He argued that

> Breathing with every step they take, wayfarers walk at once in the air and on the ground. This walking is itself a process of thinking and knowing. Thus knowledge is formed along paths of movement in the weather-world. (Ingold 2010, S121)

The statement of the empirical truth that wayfarers walk in the air as well as on the ground reminded me that the air is not only subject to the forces of the weather, but also a space of telecommunications signals, or to put it another way, the air is host to Hertzian space (Dunne and Raby 2001). A digital wayfarer walks not only along the ground in the air or in what Ingold conceptualizes as the weather-world but also traverses Hertzian space. I had cast myself as a digital wayfarer (Hjorth and Pink 2014) whose online and physical worlds are entangled in this walking mediation inspired creative process. As a digital wayfarer, I walked along paths in both the weather-world and Hertzian space, which has moorings in the ground and is of the air. My film-making process had a close and dependent relation to the ubiquitous presence of Hertzian space. Simply put, I knew it was there, and I used it as part of my creative practice. Without it, my practice would have been different.

I collated and edited my poems, posted them to my blog and asked a friend in the UK to read and record my poem. She had spontaneously made a recording of one of my poems in 2011, and I felt her voice was perfect for the kind of meditative atmosphere I wanted to conjure for this work. I was working in a way that Hamilton and Jaaniste (2009) describe as evocative research, and it was "through an ongoing dialogue between practice, theory and topic that the research question begins to make itself clear, and the shape of the research project resolves itself" (Hamilton and Jaaniste 2009, 6).

Concluding Remarks

In this chapter, I have investigated how we imagine and reimagine places and events through the use of smartphone assemblages to evoke a sense of what it's like to "be there". I have examined what part digital co-presence plays in the creation of mobile art through digital ethnography and creative practice. Co-presence and digital wayfaring have been game changers for photography and film-making. Smartphone apps have provided new ways of evoking emotional resonance.

I return to the words of Doreen Massey, who in a blog post, about a project that explored the landscape and the moving image advocated a non-representational or more-than-representational approach to film because the "stories we stumble across in this landscape are often entangled with each other, but they are autonomous too and lead off in other, unrelated, directions. There are always loose ends in space ..." (Massey 2011: n.p.).

In the next chapter, I follow some of these loose ends by picking up how poets and artists take their art into new directions by improvising and collaborating using social media to generate spaces that are both social and creative.

Note

1. http://www.ft.com/intl/cms/s/2/e56964b0-1a22-11e5-a130-2e7db721f996.html#slide0.

REFERENCES

Benjamin, Walter. 1931. *A Short History of Photography, Trans, Phil Patton.* New York, NY: Artforum 15 (6), 1977, http://www.imaginealthepeople.infor/Benjamin.pdf.

Berry, Marsha., and Margaret Hamilton. 2010. Changing Urban Spaces: Mobile Phones on Trains. *Mobilities* 5 (1): 111–129.

Berry, M. 2013. 'Being there: Poetic landscapes', in Coolbah: Special Issue, Universitat de Barcelona, Centre d'Estudis Australians, Barcelona, Spain, vol. 11, pp. 85–96.

Berry, Marsha. 2014. Filtered Smartphone Moments: Haunting Places. In *Mobile Media making in an Age of Smartphones*, ed. Marsha Berry and Max Schleser. New York: Palgrave Pivot.

Berry, M. 2016. 'Evocative moments with smartphone cameras'. In *Proceedings of the 2016 Australian Screen Production Education and Research Association (ASPERA) Annual Conference, Australian Screen Production Education and Research Association (ASPERA)*, 1–11, Sydney, Australia, (Screen Production Research: The Big Questions).

Birkerts, Sven. 1982. Walter Benjamin, Flâneur: A Flanerie. *The Iowa Review* 13 (3): 164–179. Web. Available at: http://www.ir.uiowa.edu/iowareview/vol13/iss3/42.

Bissell, D., and G. Fuller (eds.). 2011. *Stillness in a Mobile World.* London, UK: Routledge.

Bolter, D., and R. Grusin. 1999. *Remediation.* Cambridge, MA: MIT Press.

Bourriaud, N. 2002. *English Version. Relational Aesthetics.* Dijon, France: Les Presses du Reel.

Caoduro, Elena. 2014. Photo Filter Apps: Understanding Analogue Nostalgia in the New Media Ecology. *Networking Knowledge* 7 (2): 67–82.

Constine, Josh. Instagram, Technology's Window to the Soul. *Tech Crunch.* https://techcrunch.com/2013/06/20/to-thee-i-do-commend-my-watchful-soul-ere-i-let-fall-the-windows-of-mine-eyes/.

Crang, Michael. 2015. The Promises and Perils of a Digital Geohumanities. *Cultural Geographies* 22 (2): 351–360.

Derrida, J. 1993. *Specters of Marx: The State of the Debt, the Work of Mourning, & the New International, Trans. Peggy Kamuf.* London: Routledge.

Duma, Joanne. 2012. The Making of an Image: Receptivity and Transitional Space. http://www.apadivisions.org/division-31/publications/articles/british-columbia/duma.pdf.

Dunne, Anthony. 2001. Hertzian Tales: Electronic Products, Aesthetic Experience, and Critical Design MIT Press. March 2001. Pg 21, as quoted in Fried, Limor. Social Defense Mechanisms: Tools for Reclaiming Our Personal Space. MIT Media Lab Graduate Theses. Jan 2005. p. 7.

Dunne, A. and F. Raby. 2001. *Design Noir: The Secret Life of Electronic Objects.* Springer Science & Business Media.

Eco, U. 1979. *The Role of the Reader: Explorations in the Semiotics of Texts (Vol. 318).* Indiana University Press.

Eight Iconic Photographers. https://www.lomography.com/magazine/322988-iconic-photographers-who-shot-instants.

Fallon, C. 2014. Streams of the Self: The Instagram Feed as Narrative Autobiography. In *Proceedings of the Interactive Narratives, New Media & Social Engagement International Conference.*

Gadd, Nick. 2015. Circling the City. *Griffith Review* 53. https://griffithreview.com/articles/circling-the-city-mapping-forgotten-tracks/

Geertz, Clifford. 1973. *The Interpretation of Cultures: Selected Essays.* New York: Basic Books.

Geertz, Clifford. 1988. *Works and Lives: The Anthropologist as Author.* Stanford: Stanford University Press.

Goffman, Erving. 1963. *Behavior in Public Places; Notes on the Social Organization of Gatherings.* New York: The Free Press.

Gómez Cruz, Edgar, and Meyer, Eric. 2012. 'Creation and Control in the Photographic Process: iPhones and the Emerging Fifth Moment of Photography', *Photographies* 5 (2): 203–211.

Hamilton, J.G., and L.O. Jaaniste. 2009. The Effective and the Evocative: Reflecting on Practice-Led Research Approaches in Art and Design. *Interventions in the Public Domain.*

Hannan, Kevin, Mimi Sheller, and John Urry. 2006. Editorial: Mobilities, Immobilities and Moorings. *Mobilities* 1 (1): 1–22.

Hjorth, Larissa. 2016. Mobile Art: Rethinking the Intersections Between Art, User Created Content (UCC) and the Quotidian. *Mobile Media & Communication* 4 (2): 169–185.

Hjorth, Larissa, and Sarah Pink. 2014. New Visualities and the Digital Wayfarer: Reconceptualizing Camera Phone Photography and Locative Media. *Mobile Media & Communication* 2 (1): 40–57.

Ingold, T. 2010. "Footprints Through the Weather-world: Walking, Breathing, Knowing." *Journal of the Royal Anthropological Institute* (N. S.) 16: S121–S139.

Kabat-Zinn, J. 2005. *Wherever You Go There You Are.* New York: Harper Collins.

Kelly, Patrick. 2014. Slow Media Creation and the Rise of Instagram. In *Mobile Media Making in an Age of Smartphones,* ed. Marsha Berry and Max Schleser, 129–138. New York: Palgrave Pivot.

Kuchler, Hannah. 2015. http://www.ft.com/intl/cms/s/2/e56964b0-1a22-11e5-a130-2e7db721f996.html#slide0.

Massey, Doreen. 2011. Landscape/Politics/Space: An Essay. https://thefutu-reoflandscape.wordpress.com/landscapespacepolitics-an-essay/ Accessed 20 April 2016.

Miller, James. 2014. The Fourth Screen: Mediatization and the Smartphone. *Mobile Media & Communication* 2 (2): 209–226.

Picarelli, Enrica. 2015. Why Retro Looking Photography Matters for Fashion Blogging. https://afrosartorialism.wordpress.com/2015/06/08/why-retro-looking-photography-matters-for-fashion-blogging/.

Pink, Sarah. 2009. *Doing Sensory Ethnography*. London: Sage

Pink, S., and L Hjorth. 2012. Emplaced Cartographies: Reconceptualising Camera Phone Practices in an Age of Locative Media. *MIA (Media International Australia) Incorporating Culture and Policy: Quarterly Journal of Media Research and Resources* 145: 145–155.

Postill, John. 2015. Digital Ethnography: 'Being There' Physically, Remotely, Virtually and Imaginatively. http://johnpostill.com/2015/02/25/digital-ethnography-being-there-physically-remotely-virtually-and-imaginatively/. Accessed 1 April 2016.

Schleser, M., G. Wilson, and D. Keep. 2013. Small Screen and Big Screen: Mobile Film-Making in Australasia. *Ubiquity: The Journal of Pervasive Media* 2 (1–2), 118–131.

Sheller, Mimi. 2014. Mobile art. In *The Routledge companion to mobile media*, ed. G. Goggin, and L. Hjorth. 197–205. New York, NY: Routledge.

Thich, Nhat Hahn. 2015. *How To Walk*. Berkeley: Parallax Press.

Turkle, S. 2008. Always On/Always On You: The Tethered Self. In *Handbook of Mobile Communication Studies*, ed. James E. Katz. Cambridge, MA: MIT Press.

Vannini, Phillip. 2015. Non-representational Ethnography: New Ways of Animating Lifeworlds. *Cultural Geographies* 22 (2): 317–327.

White, M. 1952. The Camera Mind and Eye, Magazine of Art, 45, (January 1952). In *Minor White: A Living Remembrance* (1984), p. 36. New York: Aperature.

CHAPTER 5

Improvising and Collaborating Creatively with Social Media

INTRODUCTION

@Mimi has her Twitter account open on her smartphone. She checks her timeline and sees a haiku from @shadowX using the hashtags #haiku #haikuprompt99 #micropoetry #3lines. She sees @lost_dreamer is publicising a new blog post using the #micropoetry and #haiku hashtags. She visits @lost_dreamer's blog and leaves a comment. She loves the vivid commonplace images in his short form poems. She returns to Twitter and clicks on the #haikuprompt99. The prompt word is 'melt'. She's in the mood to play. As usual there are a group of people riffing of each other as well as people simply writing haiku following the 5-7-5-syllable convention using the prompt word. @shadowX is engaged in a witty and somewhat sarcastic debate with @Yinnocent and @bifocal about an American presidential candidate. Some of her other acquaintances are playing with juxtapositions to the notion of melting and trying to come up with bizarre images. And others have taken the erotica route with the prompt. @Mimi tweets her offering and hoped some one would pick it up and run with it.

In the previous chapter, I established how artists and writers are using social media for evocative expressions and provided a braided account of the dynamic relations between smartphone assemblages and embodied multi-sensory aspects of visual creative practices with smartphone cameras. Here, as the title of this chapter suggests, I delve into creative practices with poetry that we can readily observe in social media ecologies but before I do this, I would like to refer back to a quote I used

© The Author(s) 2017
M. Berry, *Creating with Mobile Media*,
DOI 10.1007/978-3-319-65316-7_5

87

in Chap. 1 from Tim Ingold in his foreword to Vannini's edited collection of essays called *Non-Representational Methodologies: Re-Envisioning Research*. Ingold urges us to meet the world by saying "Enough of words, let's meet the world" (Ingold 2015, vii). In this chapter, we will meet the world through looking at how people use poetic forms to improvise and collaborate with each other. I am not concerned with literary analysis or the representational dimensions of poetry here; rather, I am interested in how short form poems are a way of meeting the world to share lifeworlds across national boundaries and time zones through social media hashtags and special interest groups, which may be conceived of as communities of creative practice. I am also interested in how relational aesthetics (Bourriaud 2002) come into play when artists and writers begin to improvise and collaborate in creative community spaces enabled by the functionalities of mobile devices such as smartphones and by social media. I also propose that relational aesthetics in these mobile media contexts are framed by a sense of co-presence that spans time zones, geographic distance and synchronicity.

Obviously, social media has generated substantial challenges to traditional media and publishing outlets in terms of both production and consumption of content according to Leadbeater (2009) because the media and publishing industries no longer control the flow of information or whether or not particular content is to be published. The consequences of this are far-reaching and largely beyond the scope of this chapter. What is of importance here is that social media ecologies have provided writers with environments where innovative forms of writing and ways of working flourish. For example, in Twitter, socialities emerge around hashtags, which may be conceived of as interconnected ecosystems within larger media ecologies. Intrinsic to such social media ecologies is the notion of digital co-presence whereby there is a tacit assumption that people do notice and pay attention to each other's posts. I provided a discussion of digital co-presence how it can provide seeds for new forms of mobile art practices in Chap. 4. Improvising and collaborating with others is a notable feature of how artists and writers use these social media spaces for their creative practice.

These socialities and emplaced visualities may be understood as performance or performative in function. Hashtag streams rely on participation and improvisation. There is an oscillation between people, poems, images and words in Twitter hashtag streams. If I choose to join, my hope is that my poetic tweet will be "retweeted" by others to appear further along in the stream and that followers will respond with a comment or a verse of their own. Jumping into a Twitter stream of a haiku, tanka, gogyohka (contemporary Japanese short form poetry akin to tanka) or senryu (Japanese short form wry self-referential poetry akin to haiku) hashtag as presented in the opening vignette is an act of improvisation and dialogue as well as a practice and performance.

Facebook also offers a platform where dedicated spaces for creative practice may be set up and where digital co-presence is assumed. Public social media creative writing groups such as Haiku International[1] and Twitter hashtags such as #haiku give rise to creative spaces where people gather to play with words. These hashtags timelines and Facebook pages may be conceptualized as theatrical stages where the play has no script and no apparent structure where sometimes group collaborations happen quite spontaneously. At other times, they may be loosely pre-planned to accommodate different time zones or take on asynchronous format. Social media offers creative writers ways of participating and collaborating in global and mobile social media spaces where writing takes on a performance dimension and where performers are aware of each other's digital co-presence. Their posts are performative (Austin 1955) as well as performance (Schechner 1985, 2002). There is a dramaturgy at play framed by relational aesthetics.

POETRY AS PARTICIPATORY PUBLIC ART

In these spaces, short form poems become forms of public art that is dependent on relational aesthetics as expounded by Bourriaud (2002) where the artist "catches the world on the move" (Bourriaud 2002, 14) so that each poetic offering becomes "a proposal to live in a shared world" (22). Aesthetics which are shared in hashtags and Facebook pages dedicated to poetry including the use of rhythm; tropes including metaphors, metonymies and personification; schemes ordering words and syntax, including unconventional use of normal speech; and lines of text broken up according to the number of syllables (Abrams and Harpham 2005) and these have been remediated and adapted for mobile media contexts. Poetry uses "similarities in form and position among certain words in the text, similarities that are rationalized and interpreted in terms of meaning" (Riffaterre 1983, 36). Poetic language works through systems of descriptions so that the semantics of a text invite the reader to engage in a process of accumulation to ascribe meaning and to draw syntagmatic connections or associations between configurations and juxtapositions of words (Riffaterre 1983). These features of poetic language have vitality and contribute an animated dynamism to Twitter hashtag streams and Facebook pages dedicated to poetry.

Interactions in the digital creative spaces I have observed and participated in as an artist and writer since 2010 are bound by emergent, dynamic and tacit rules of engagement that quickly become normative conventions of exchange, challenges, ripostes and even gift giving in the form of emoticons, and pictures of flowers and cute animals. There is an implicit "sense of honour" (Bourdieu 1977: 15) tweeting and retweeting amongst poets and artists. A "sense of honour" is intrinsic to a community's logic of practice according to Bourdieu (1977) where by it is "...the cultivated disposition, inscribed in the body schema and in the schemes of thought, which enables each agent to engender all the practices consistent with challenge and riposte..." (Bourdieu 1977: 15). A sense of honour maintains harmonious and respectful relations within a community while at the same time accommodating rigorous debates with opposing points of view. People who troll are simply blocked and reported to administrators by participants.

A sense of honour as well as relational aesthetics is entangled into the fabric of poetry hashtags where each offering is a performance awaiting further unfoldings (Schechner 2002). Some will bring disruptions; some will be variations on a theme that grows into a ritual. An example of how variations on a theme can become ritualized is the background history of a Twitter hashtag called #theremustbecrowsinit. A successful poet who had a haiku knocked back by a prominent journal initially started the hashtag in 2010. The reason given to her for the rejection by the editors of the journal was that they had already published several haiku referring to crows and didn't want any more. She tweeted her chagrin in a humorous way about this rejection using this hashtag. Her circle of followers rallied around and began posting short form poems with crows using the same hashtag. This hashtag still continues at the time of writing—2017, and it is also an example of how an artist or poet can "set his [her] sights more and more clearly on the relations that his [her] work will create among his [her] public, and on the invention of models of sociability" (Bourriaud 2002, 28). The poet was annoyed with a rejection slip and decided to use this to create a new work knowing that the conviviality of poetry hashtag spaces would result in other writers taking up her call that there will never be too many haikus about crows. Indeed, one could claim that each use of the hashtag #theremustbecrowsinit is a convivial wink to the creator of the hashtag by those who know the backstory.

The nature of the occasion, the genre and an awareness of the audience are all important in poetry hashtag streams. Allsop notes "the conventionalized (and therefore often unquestioned) relations between writing and performance are proving increasingly inadequate as interdisciplinary and cross-disciplinary arts practices emerge in response to rapidly shifting cultures" (1999: 76). As I noted in earlier work (Berry 2011), the act of writing takes performance dimensions as well as communication and self-expression within Twitter contexts. People don't always use their real names. Pseudonym user names are quite commonplace. I observed that people do not ask conventional questions designed to elicit personal information as they would at a face-to-face social gathering such as a conference or a party. Instead, there is a tacit understanding that one does not ask questions concerning the place of employment or study, occupation, marital status, etc., in the public stream. (Twitter has included a private one-to-one direct message function for such personal questions and for more intimate interactions.) Nevertheless, close friendships and easy social relations develop between people and spill into the physical world.

Literary forms and aesthetics are remediated (Bolter and Grusin 1999) in mobile and social media, and new forms are evolving in these ecologies. In the late 1990s in Japan, the novel form was remediated so that people could read novels on mobile phones using the SMS functionality. The Japanese referred to this new form as *keitai*. This form was criticized as adolescent, shallow and genre based by Jane Sullivan (2008, 28) who said that Japanese mobile phone novels suffer from "quality control". On the other hand, Coates observed *keitai shōsetsu* stretch reading as well as writing because the "unusual orthographical features found in *keitai shōsetsu*, focusing on emoticons, symbols, non-standard punctuation and unconventional script choice and size, resulting in a novel which is appealing and easily accessible to young people" (2010, 1). Furthermore, *keitai shōsetsu* have their origins in "high" Japanese literature such as Murasaki's *The Tale of Genji*. They incorporate the poetics of traditional Japanese literary canons so that "*keitai shōsetsu* must be conceptualized as an extension of older media practices and epistolary traditions" and further they "invoke the art of *haiku* poetry" (Hjorth 2009, 35). In short, mobile phone novels are examples of mobile media art that remediate established forms and aesthetics in this case they remediate and seed hybrid Japanese literary forms into mobile media ecologies to expand creative practices. It would appear that Sullivan (2008) overlooked the complexity of *keitai shōsetsu* when she arrived at her judgement that there is a lack of quality and depth in this new form.

This point about the complexity of mobile media forms was picked up Raley (2009), in a paper presented at the Digital Arts and Culture Conference where she argued that mobile media performances have literary uses and merit. She cites SMS poetry contests as her primary evidence and asks: "can anything of value emerge from a medium whose principle literary genre is the adolescent romance (mobile phone novels) or from a platform that promotes instantaneity and ephemerality (Twitter)?" (1). She proposes that if we focus on the processes of making poetry in mobile media ecologies we can begin to appreciate its merit. She explores the role of improvisation in relation to intersubjectivity and relational aesthetics based on "Canetti's sociological investigation crowd dynamics" (6) and argues that "barriers separating self and the crowd dissolve in a process not of assimilation but of temporary and incomplete transport" (6) so that poetic practices in mobile media contexts are more about sharing intersubjective aesthetics than creating independent symbolic expressions. In other words, poetry becomes a social activity and goes beyond representational dimensions in mobile media ecologies.

Neil Gaiman was inspired enough by the potential value of Twitter as a mobile media performance space to compose an audiobook in 2009 called *Hearts, Keys and Puppetry* which commenced with this tweet:

> Sam was brushing her hair when the girl in the mirror put down the hairbrush, smiled & said, "We don't love you anymore." @BBCAA #bbcawdio

Other Twitter users were invited to join and the final version was complied in reverse order by BBC Audiobooks America. Over a hundred people contributed to this collaboration. The potential of social mobile media as an environment for improvisation and collaboration has been increasingly recognized and utilized by poetry circles. There have been notable British poets according to Charlotte Cripps (2013) who have experimented with Twitter as a way of writing and sharing poetry. These include Benjamin Zephaniah, winner of the T.S. Elliot prize, George Szirtes, Ian Duhig, Sophie Robinson, Andrew McMillan, Inua Ellams and Alison Brackenbury amongst others. Cripps observed, "collaborative poetry projects are becoming popular too" (2013, n.p.). @Mimi's practices with Twitter are not uncommon. In the next section, I present further insights from my research.

Rehearsed and Impromptu

In 2010, I embarked on a digital ethnography of Twitter hashtag time-lines. I participated both as an observer as well as a creative practitioner. This provided me with a window into the processes of making poetry. I asked a question about whether people wrote their poems straight into Twitter or whether they crafted their offerings elsewhere to gain a sense of how many rehearsed their poetry performances and how many would simply jump in impromptu. Unsurprisingly, responses I received were varied and like in many other creative writing contexts, there are those who carefully work through numerous drafts planning each Tweet and those who work more intuitively and like to leap into the flow to see what will unfold. My findings resonate with George. R.R. Martin's famous distinction of two types of writers:

> I think there are two types of writers, the architects and the gardeners. The architects plan everything ahead of time, like an architect building a house. They know how many rooms are going to be in the house, what kind of roof they're going to have, where the wires are going to run, what kind of plumbing there's going to be. They have the whole thing designed and blueprinted out before they even nail the first board up. The gardeners dig a hole, drop in a seed and water it. They kind of know what seed it is, they know if planted a fantasy seed or mystery seed or whatever. But as the plant comes up and they water it, they don't know how many branches it's going to have, they find out as it grows. And I'm much more a gardener than an architect. (Martin 2011)

But no matter whether my respondents identified as "architects" who like to plan or gardeners who like to plant seeds and let them grow, all agreed that there was an element of performance in their tweets and that they were playing to an audience of co-present others. When thinking about performance within the expanded field of online social life and life-worlds, it is still relevant to consider representation of the self and performance personas, simulation, presence and the subjunctive "what if" as well as the elusive indicative "as if" (Schechner 2002). Schechner, one of the founders of performance studies, provides a conceptual lens through which this material and observable phenomenon may be ana-lysed. Schechner (2002) makes connections between actor training and Victor Turner's theory of social drama and reworks Turner's notion of the liminal[2] into a conceptualization of the theatre space as liminoid to resembling Turner's liminal space, which is predicated on a willingness to suspend judgement and disbelief so that people, spaces and objects are layered with symbolic meaning and theatre aesthetics. Turner's (1979) concept liminality refers to "a state or process which is betwixt-and-between the normal, day-to-day cultural and social states and processes of getting and spending, preserving law and order, and registering structural status" (465).

Performances in mobile and social media have parallels with the process an actor undergoes to create a liminoid state of "not not me" (Schechner 1985) when entering a rehearsal or performance space to create a three-dimensional performance text. In theatre and performance art selves move between the present grounded here and now and imaginary worlds. Lived experience zigzags between these states as "performance 'takes place' in the not me... not not me" between performers; between performers, texts and environment; between performers, texts, environments and audience (Schechner 1985: 113). The Twitter space provides selves with a threshold space in which to play and to explore the relationships of textually inscribed personas and bodies. People experiment with personas, which may be regarded as performances. I applied this lens to my observations of self-identified poets playing in Twitter spaces defined by hashtags.

I noticed that many like to take on archetypes famously identified by Jung, to shape performances of their online personas such as the everyman or regular girl, the hero or heroine, the innocent, the lover and the wise mentor. Some treat Twitter as a play space where they can be "the joker in the deck" through engaging in "the supreme *bricoleur* frail transient constructions, like a caddis worm or a magpie's nest " (Schechner 2002: 80) where much behaviour involves subjunctive spaces and restored behaviour. Restoration is a term Schechner used to refer the main characteristic of performance where living behaviour is "treated as a film director treats a strip of film" (Schechner 1985: 35). Behaviour may be put on in much the same way as a mask and is always subject to revision through a rehearsal process. It is symbolic and reflexive so that the self can act as if it were another is "twice-behaved behavior" (Schechner 1985: 36) in that it can be worked on and changed. Twitter may be imagined as a stage where the play has no script and no apparent structure where sometimes group collaborations happen quite spontaneously. There are ongoing public flirtations as well erudite debates on geopolitics and philosophy. There are entrances and exits with lines delivered by players with names such as bookwriter222; where continuums operate in terms of how various selves present themselves to themselves and to others, and how they revise their performances over time. Others with more of a desire to play the regular guy improvise with haiku all day using their daily rituals and social activities as grist for their mills of poetic inspiration. Below is a fictionalized vignette based on my observations to show how this might transpire:

> An elderly Chinese man called Luke is in a village somewhere in South East Asia. He is a Christian pastor and lives there with his wife. He loves parables and believes in the power of simple stories. He likes to lead by example and show people that he is just an ordinary guy who does ordinary things – a typical solid citizen. He has found social media is a good way to reach out to people outside of his parish and spread the message in a subtle way. Every day except Sunday he looks after his grandchildren with his wife while the parents go to work. A family of cats lives in their garden that keeps the rats at bay. He watches his grandchildren play with the cats in the garden. The cats are chasing paper butterflies they made together earlier out of an old newspaper. He writes a haiku about cats chasing butterflies that are not real and posts it to Twitter using the hashtag #butterfliesRfree. Within minutes it is retweeted several times. @foxclown01 writes a gently mocking response using the image of butterflies and paper. He smiles – @foxclown01 always has to make a joke and usually at his expense. He writes a rejoinder in the form of a self-deprecating haiku. In the meantime, other poets have jumped into the fray and are tweeting haiku about butterflies. A chain of butterfly haikus has commenced using his hashtag. He writes a sequel where the cat does catch a real butterfly and is surprised. He will make the haiku Twitter stream into a story to tell the children later about butterflies who see all the wonders of the world, as they follow the sun in a hot air balloon bringing love and peace to all the places they visit.

> Now it's lunchtime – his wife has prepared a salad. He writes a short poem about gratitude and the bountiful colours of the herbs and vegetables. Again it's retweeted and others write about food as well. He sends flower scented hugs to all those who retweet his poems. He feels he's made so many new friends who share his love of haiku. They've become familiar faces and he enjoys their presence in his life everyday.

Twitter and Facebook poetry circles have provided Luke with a way of crossing borders to form new relationships. He puts forward his best face and is concerned to come across as an ordinary grandfather than a pastor and encourages people to chat and engage in banter. Luke's poetry and online interactions are a performance yet at the same time they reflect who he is. More recently, in *Performed Imaginaries* (2015), Schechner reiterated how he conceptualizes performance by saying that "As I theorize it, something g 'is' performance when according to the conventions, common usages, and / or traditions of a specific culture or social unit at a given historical time, an action or event is called a 'performance'" (Schechner 2015, 6). He proposes a manifesto of four principles in an impassioned and idealistic plea for a "New Third World of Performance" that he feels may help "save the world" (2015, 1).

1. To perform is to explore, to play, to experiment with new relationships.
2. To perform is to cross borders. These borders are not only geographical, but emotional, ideological, political and personal.
3. To perform is to engage in lifelong active study. To grasp every book as a script—something to be played with, interpreted, and reformed/remade.
4. To perform is to become someone else and yourself at the same time. To empathize, react, grow, and change (Schechner 2015, 9).

This manifesto can be adapted for performance in social media spaces where people don personas that improvise and experiment and form new and collaborative relationships that transcend geography and time zones. It can also be used as a rubric through which to examine socialities and emplaced visualities in mobile and social media. In the next section, I present two case studies of people improvising and collaborating creatively with mobile and social media.

A Chained Poem

In my digital ethnography, I've found that self-identified creative writers enjoy extending poetic forms through improvisation and collaboration. Co-presence and intersubjective relational aesthetics are both crucial for these mobile media creative writing practices. To join a particular poetry hashtag stream implies an awareness of other performers, texts, audience as well as one's own subject position within a writing act. One also needs to enter threshold spaces that Schechner (2015) would call "New Third World of Performance" spaces where borders like time zones and geography cease to matter. In 2010–2011, I was involved with many such performances where threshold spaces were created with a hashtag and a network of poets would gather to bounce short forms poems together. We became familiar with each other's other work and developed friendly relations using Twitter to talk to each other about our poetry and other everyday things. We'd look out for each other in poetry hashtag streams. Some of these hashtags included a prompt word that needed to be used in poems with the hashtag.

#*Haikuchallenge* is an example of one such hashtag stream. Each day poets use the hashtag to post their haiku containing a specific word. One of the regular participants will prescribe a word to be used in haiku for that day. This hashtag has remained active into 2017. I met numerous poets in these hashtag streams. We began to leave haiku as tokens for each other and look for each other's presence. Some days we'd play with flowers, other days with mythical beasts such as dragons, unicorns and vampires. We cross into imaginary worlds with our personas and improvise loosely connected chains just like in the vignette in the previous section. For the poets in my circles, improvisation and collaboration included inhabiting strange bricoleurs of transient and ephemeral structures in order to create and restore subjunctive spaces. We often paid homage to the origins of haiku aesthetics by creating imaginary courtly Japanese houses with elegant gardens and ponds as *bricoleur* structures, which may be restored and inhabited through a performance within a hashtag poetry stream. We became intrigued with classical Japanese poetry and stumbled across renga in our search for more knowledge about forms and structures.

Collectively, we Googled to find out more and discovered that renga is a literary art form written by two or more collaborators that was popular in Japan during the medieval and Edo periods. Bashō was a keen proponent and innovator of the form. As well as being poetry, renga is an improvisation where two or more poets collaborate to produce a performance through linked verse. Its popularity declined during the Meiji period when Japanese poetry became influenced by Western forms where the emphasis was on the work of an individual poet rather than on collaborations. Reichhold (1995) describes the writing of renga thus,

> While engaged in contributing to a renga the writers are aware that a careless, uneducated reader could find the completed poem merely a crazy-quilt assemblage of alternating two and three line stanzas. However, they also know that a knowledgeable, sensitive reader will legitimately anticipate

a creative, meaningful and enjoyable assemblage of organized stanzas. Writers of renga design their lines so that ANY ADJACENT SET of two and three lines will have a subtle but recognizable relationship.

She offers the following advice to readers:

> As you read some of the renga the important thing to watch is what happens BETWEEN the links. Think of each stanza as a springboard from which you are going to jump. As your mind leaps (and you think you know where the poem is going) you should be forced to make a somersault in order to land upright in the next link. It is the twist your mind makes between links that makes renga interesting.

Renga takes place in the relations between performers, text, environments and audience. A writer participant in a renga must attend to drawing syntagmatic connections with the preceding verse as well as leaving opening for the writer of the next verse. In other words, a relational aesthetics is embedded within the act of writing.

We had been sharing material about renga forms over the past couple of days and I remember @bookwriter22 posted a link to a Wikipedia article. We all read the article and decided we really should have a go at writing one so that we could learn more about this form through experimenting with it ourselves. We all already had an established haiku practice so it was really a matter of extending our respective practices into a new and unknown area for us using mobile and social media. @peterwilkins1 and I said we would like to capture the verses and post them up on our respective blogs to document the experience. I also asked the group if I could write about this in academic journals as a creative practice-based experiment as well. Everyone willing agreed to the renga being documented the two blogs and also that I could write about the experience in academic forums.

In January 2011, we chose a day and a time for the renga and decided on the order in which each of us would write and set the commencement time using Greenwich Mean Time. We talked about our planned event amongst ourselves as a way to letting our followers know that we were going to perform an improvised poetry event. @marousia (myself), @peterwilkins1, @bookwriter222, @amoz1939 and @remittancegirl were the renga performers. We followed the conventions outlined in *wikipedia* for Kasen renga. The conventions follow that of a polite social gathering where alcohol is served. Normally, a renga opens with small talk about the weather and references to flowers and nature, the middle part is loose where various themes such as love, religion and politics are discussed, the last section may become disjointed and a bit raucous as the party winds up. Traditionally, renga forms have this syllabic pattern, 5/7/5, 7/7, 5/7/5, 7/7 and so on. The 7/7 verses may be thought of as a clap.

We marked each verse with the hashtag #rengachange so that we could keep track of the verses as well as the turn taking. The poem was composed over two days to accommodate time zone constraints. We also anticipated that our followers would see it, hopefully become intrigued and follow the hashtag #rengachange. Two of the renga performers were in the UK, two were in South-East Asia and I was in Melbourne, Australia. The opening two verses of the renga, which we later called *Fluxus Interruptus* show a conventional opening with references to the weather and nature. The second stanza employs a meme in the form of cherry blossom, which symbolizes a Zen aesthetic:

1. Under a cool moon/ the earth slumbers, breathes softly/ stillness and shadows
2. A gentle western breeze lifts / petals from a cherry tree
   The poem continues in this polite vein
3. Blue cat very still/ watches butterflies/ admid the petals
4. Dragonflies smiling -- butterflies preen in colors ~ having my sake
5. Silver ripples on green pond/ shades of orange, Koi resting
6. All is quiet here / pondering the coming day / nothing stirs, I sit
7. Ginko trees stand silent guard/ maples bowing to the moon
8. Bamboo house clean and ready / sake bowls fill to the brim
9. Distant thunderstorm/ heron glides down mountainside/ soft patter of rain

Until a stranger arrives about one-third of the way through to disrupt the atmosphere:

10. A stranger arrives from town / the thunderstorm is nearing
11. horizonal rain / steals the stranger's hat and laughs / gifts it to the wind
12. The hat staggers drunkenly/ spins wildly on the pond's brink
13. Frog jumps out of pond ~ and landing on floating hat ~ a nice lily pad
14. Stranger kneels beside the pond/ Koi feeding upon his smile

15. As dawn breaks cover / wrathful angry storm abates / I wait patiently
16. Air fresh and dense with thunder / dew-trapped sparks of lightning passed
17. Fire gives out warmth ~ all guests seated by crickets ~ thunder storm ceases

A metaphor of a dinner party is reasserted as the sake flows. This line also signals that it is time to make the links between verses quite loose.

18. Feast of eels and sake/ appears for the dinner guests
19. Crickets one sake ~ waiting for a guest to come ~ eels are delicious
20. Violet mist drifting down/ spreading blanket over guests
21. As frogs leap and sing / spoken tales of woven dreams / synchronicity
22. No leaping frogs, no grilled eels / stops the slow drip - spilt sake
23. Old man sings a song/ of brave warriors, their swords/ and lovely geishas
24. Sun climbs high, the morning burns/ sweat trickles down guests' faces
25. The mist clears away / a breeze that whispers secrets / cools the troubled mind
26. Breath upon a dusty glass / the ghostly geisha's kiss print
27. Ghostly foot falls tap/ across the sun light terrace/ mosquitoes buzzing

The party begins to draw to a close with some guests drunkenly dozing, others flirting and some retiring for the night:

28. Sated guests lie on futons/ soothed by ghostly lullabies
29. An exchange, a glance / a frisson of excitement / quivers up the spine
30. In the corner the koto / sounds one dark abandoned note
31. Kimonos rustle/ the sound of slapping rhythms/ paper screen snaps shut
32. Far away, behind mountains/ black ships in turquoise ocean
33. Surging through the waves / a promise of a future / ships that speak of change
34. A floating island sitting / atop a giant turtle
35. A demon appears/ head flaming, eyes fiery/ the gods are smiling

In the final verse, the guests depart, returning to their own lives:

36. With lightening speed, he strikes / the gods of change start
dancing

Peter's blog post for *Fluxus Interruptus* (2011) provided an accurate description of the process of writing the renga live in Twitter. In it, he acknowledged a debt to Dada as well as to Japanese aesthetics:

> Despite our moments of confusion the predominant feeling that ran the whole way through this exercise was one of great fun. It did not seem to matter to any of us if we strayed from accepted procedure or if our verses lacked continuity. On the contrary, we were poetic rebels & celebrated the fact that a degree of Dadaism had crept into our work as we 'laughed in the face of order & convention.'[3]

It's clear from this blog entry that relational aesthetics came into play in this collaborative improvisation. The improvisation was way of meeting the world to share lifeworlds across national boundaries and time zones through social media hashtags and special interest groups, and the group of poets who came together to collaborate is a community of creative practice that transcends national and time zone borders.

A PARTICIPATORY MOBILE MEDIA ART PROJECT

The next project I present is *Twitter Poetry #conku*—a participatory mobile media art project. As I had already established through my ethnographic and sometimes auto-ethnographic research, Twitter offers creative writers ways of participating and collaborating in global online poetry communities where writing takes on a performance as well as a social dimension. In this particular project, I brought poets together to perform poetry with Twitter in an exhibition space blending impromptu poetry with open mic, performance and mobile media art. The proposition behind *Twitter Poetry #conku* was that Twitter and other social media are ready to be appropriated and colonized for improvised participatory poetic performances with both words and visual images. As such this project can be considered an example of intentional creative misuse as espoused by Farman (2014). It was part of a larger exhibition designed to showcase creative practice and design research at RMIT University (Fig. 5.1).

between
two hands
creation claps

Fig. 5.1 From *Twitter Poetry #conku* project. Photograph and text Marsha Berry

I was provided with an exhibition space for the *Twitter Poetry* performance. It was a white room gallery space. I had the space prepared with stools placed in a circle, the projector, screen and a microphone for reading the haiku. I connected a laptop to a projector and projected a live Twitter feed onto a large screen. The poets sat on stools in the centre of the room facing the screen so that they could see the Twitter timeline and respond. They used the hashtag #conku and composed free form haiku. I had told my own community of practice and writing buddies on Twitter about this event in advance and invited them to join if in if they felt inclined. The poets in the exhibition space took it in turn to read out the poems in the live feed. Some wanted to read out their own poems open microphone style while others were happier for someone else to read out their poems. The audience was able to hear the poems as well as see a live feed of these poems mingling with poetry from around the globe in a stream of tweets. If audience members wished, they could take up the invitation to participate in the live feed and quite a few did. Below is a narrative description of the event based on my field notes. The poems are taken from the #conku hashtag stream. The hashtag itself is an amalgamation of two words—"convergence" as a nod to the overall exhibition and "haiku" to signify the type of poetry being performed.

The space itself had grids of circular black dots on the white walls. This was a design motif that ran through the whole of the building. The poet participants began to arrive. Some I already knew and others had responded to a call for poets through the newsletters of various organizations for creative writers such as Australian Poetry. I made any necessary introductions and we talked about types of haiku and Twitter. I made sure everyone could login to the gallery space Wi-fi and could post to Twitter using the #conku hashtag. We could see our tweets on the wall projection. One of the poets began to read the tweets using the microphone. I took a photograph of the space and posted it saying, "twitter poetry meets open mic". Then a tweet came into the stream from one of my British connections—he wanted to know what we thought the collective noun for poets was and could we answer in haiku? The game was on. Below is one of the responses from the poets in the gallery space:

Mel Jepson @MelJepson
What is the collective noun for poets???
What is the collective noun for cats chasing mice???
#conku #haiku

Then attention inside the exhibition space turned to the room itself because it was quite an unusual space dominated by the dots in a grid. Below is a selection of poetry tweets that show how the poets in the space were riffing off each other to create an improvised flow in reaction to the design motif of the black dots in grids. At this juncture in the performance, the poets were focused on each other's tweets while people wandered in and interacted with the poets, the live Twitter feed and listened to the open mic style performances. The digitally co-present audience receded into the background and were not a central concern for the performers in the room. People who were not physically co-present would join and stay for a while but the poets in the exhibition space did not pick up their contributions as readily.

Stefan Schutt @stefan_schutt
stonehenge of points
pickup in the middle

beam me up
#conku #haiku #poetry #senryu

Jenny Weight @geniwate
@maz_abroad The pattern falls apart - but the middle is holding.
#conku

Mel Jepson @MelJepson
Circular trams, circular arguments / Holes you could drive a tram
thrugh #conku

Stefan Schutt @stefan_schutt
round in square
big big spaces
floating control
to the nerve centre
#conku #haiku #senryu #poetry

Jenny Weight @geniwate
@MelJepson Surrounded by circles there may well be something behind
them ... Ah my friend the dot #conku #backtowhereibegan

Mel Jepson @MelJepson
Nerve centre of planets, shining love in the morning sky #conku

Jenny Weight @geniwate
Inside the dot lies darkness. #iwanttosee #conku

Mel Jepson @MelJepson
Such a little dot, to cause so much fuss / I guess that's just us #conku

Benjamin Solah @benjaminsolah
@maz_abroad @stefan_schutt zero-zero-zero / like little dots #conku

Marion E. Piper @maz_abroad
@marousia Connect the dots / to colour / in between the minds. #conku

marousia @marousia

@maz_abroad #conku dots of laughter / children laugh / her skirt has
dots #haiku

lucindastrahan @lucindastrahan
Dot-dot-dash / SOS / over-and-out #conku

Stefan Schutt @stefan_schutt
twinkle twinkle little dot
may you not be full of snot #conku #haiku #senryu #poetry

Marion E. Piper @maz_abroad
@benjaminsolah @stefan_schutt Poetry / is the cream / in the Oreo.
#conku

Benjamin Solah @benjaminsolah
@stefan_schutt feed the kids oreos / don't teach them poetry / they say
it's dangerous #conku

Marion E. Piper @maz_abroad
@benjaminsolah Zero-zero-zero / hero / Oreo / biscuit and milk.
#conku

Smartphones were invisibly woven into the fabric of the performance.
The digital and the face-to-face, the online and offline were entangled
together. The performance with its easy repartee between participants
went for two hours in a similar vein. Interestingly, they played to each
other and to the audience who came into the exhibition space and did
not pay much attention to digitally co-present others.

CONCLUDING REMARKS: CREATIVE PRACTICE THROUGH IMPROVISATION AND COLLABORATION

I have previously argued that the digital co-presence and the extreme
accessibility of smartphones (Berry 2016) foster the desire to share evocative moments and creative expressions. In this chapter, I have shown
how these assist improvisation and collaborations in participatory mobile
media art. The tacit rules of engagement for social media and the need
to straddle time zones in the renga project meant that social conventions
and norms governing other interactions in poetry hashtag timelines also

applied and that the time frame for a renga "dinner party" was stretched into two days to accommodate the geographic locations and time zones of the participants. Courtesies like thanking people for retweets were observed and we understood that our fellow participants would dip in and out to eat, sleep and attend to their daily routines. There were numerous posts using the #rengachange hashtag that were not poetry but rather phatic or procedural in function. Poetry and the mundane became entangled where the audience joined in from the sidelines to ask questions, offer advice and generally be part of the overall performance.

Twitter Poetry #conku was far more contained. The mundane did not really spill into the #conku hashtag. There was interest expressed from those digitally co-present but outside of the exhibition space; however, the poets in the room did not really seem to want engage beyond the physical performance space. They were more concerned to banter with each other and the audience who came into the room. The poet performers who participated had not used Twitter previously as a vehicle for creative practice and were more used to poetry open mic nights so the *Twitter Poetry #conku* project took them into unknown terrain. While they had an awareness of digitally co-present others and could see them in the live feed, they were more concerned with the liveliness of the performance space and with each other as audience.

Improvisation with all its messiness is reliant on relational aesthetics where poets have a shared intersubjective notion of how live participatory poetry collaboration may play out. In this chapter, I have presented two creative projects where words did become a way of meeting the world and presented poets with a way of being in the world to perform across borders and to play with new relationships in collaborative poetry chains. The performance qualities of writing in social media ecologies such as Twitter was discussed by the poet performers of both the renga, which was a purely online event, and *Twitter Poetry #conku* performance, which had a physical as well as online audience. What both projects had in common was a mobile media ecology that has changed conditions of possibility for improvisation and collaboration. The participants had a self-reflexive sense of being a live performer in both projects, through the extreme accessibility of mobile media the poet participants felt the aura of an immediacy through a connection with an ever-present digitally co-present audience. The success of each of the projects was dependent on relational aesthetics where "the work of every artist is a bundle of relations with the world giving rise to other relations, and so on and so forth, ad infinitum" (Bourriaud 2002, 22).

NOTES

1. https://www.facebook.com/groups/haikuinternational/?fref=nf.
2. See http://actlab.us/death/TurnerFrameFlowReflection.pdf.
3. http://peterwilkin1.blogspot.com/2011/01/fluxus-interruptus-when-renga-joined.html.

REFERENCES

Abrams, Meyer Howard, and Harpham, Geoffrey Galt. 2005/2009. A Glossary of Literary Terms, 9th ed. Boston: Wadsworth Cengage Learning.

Allsop, Ric. 1999. Performance Writing. *PAJ: A Journal of Performance and Art* 21 (1), Jan 1999.

Austin, J., L. 1955. *How to do things with words*. Oxford: Oxford University Press.

Berry, M. 2011. 'Poetic tweets'. In *Text Journal of Writing and Writing Courses, Australian Association of Writing Programs, Australia*. 15 (2): 1–13. ISSN: 1327–9556.

Berry, M. 2016. 'Evocative moments with smartphone cameras'. In *Proceedings of the 2016 Australian Screen Production Education and Research Association (ASPERA) Annual Conference, Australian Screen Production Education and Research Association (ASPERA)*, Sydney, Australia, 1–11. (Screen Production Research: The Big Questions).

Bolter, J, David and Grushin, Richard, A. 1999. *Remediation*. Cambridge MA: MIT Press.

Bourdieu, Pierre. 1977. *Outline of the Theory of Practice*. Cambridge: Cambridge University Press.

Bourriaud, N. 2002. *Relational Aesthetics*, trans. Pleasance, S. and Woods, F. Les presses du reel.

Coates, Stephanie. 2010. The Language Of Mobile Phone Novels: Japanese Youth, Media Language And Communicative Practice. In *18th Biennial Conference of the Asian Studies Association of Australia in Adelaide*, 5–8, July 2010. http://asaa.asn.au/ASAA2010/papers/Coates-Stephanie.pdf. Accessed 10 April 2011.

Cripps, Charlotte. 2013. Twihaiku? Micropoetry? The rise of Twitter poetry. *Independent*. http://www.independent.co.uk/arts-entertainment/books/features/twihaiku-micropoetry-the-rise-of-twitter-poetry-8711637.html.

Farman, Jason. 2014. "Site Specificity, Pervasive Computing, and the Reading Interface". In *The Mobile Story: Narrative Practices with Locative Technologies*, ed. Jason Farman, 3–16. New York: Routledge.

Hjorth, Larissa. 2009. Cartographies of the Mobile: The Personal as Political. *Communication, Politics and Culture* 42: 2.

Ingold, Tim. 2015. 'Foreword'. In *Non-representational Methodologies: Re-envisioning Research*, ed. Phillip Vannini, pp. vii–viii. New York and London: Routledge.

Leadbeater, Charles. 2009. We Think: The Power of Mass Creativity. http://iatelevision.blogspot.com/2009/02/charles-leadbeater-we-think-power-of.html. Accessed 5 May 2011.

Martin, George, R.R. 2011. https://www.goodreads.com/quotes/749309-i-think-there-are-two-types-of-writers-the-architects.

Neil, Gaiman. 2009. *Hearts, Keys and Puppetry*. http://www.sffaudio.com/?p=13878. Accessed 4 May 2011.

Raley, Rita. 2009. Mobile Media Poetics. In *UC Irvine: Digital Arts and Culture 2009*. Retrieved from: http://www.escholarship.org/uc/item/01x5v98g. Accessed 1 Nov 2010.

Reichhold, Jane. 1995. Renga http://www.ahapoetry.com/r_info.htm#start. Accessed 25 May 2011.

Riffaterre, Michael. 1983. *Text Production*. New York: Columbia University Press.

Schechner, Richard. 1985. *Between Theater and Anthropology*. Philadelphia: University of Pennsylvania Press.

Schechner, Richard. 2002. *Performance Studies: An Introduction*. London: Routledge.

Schechner, Richard. 2015. *Performed Imaginaries*. Abingdon, UK: Routledge.

Turner, Victor. 1979. Frame, Flow and Reflection: Ritual and Drama as Public Liminality. *Japanese Journal of Religious Studies* 6 (4): 465–499.

Sullivan, Jane. 2008. 'Thumbs Down to the Nokia Novel'. *The Age*, A3 section, 9 Feb 2008, p. 28.

Wilkin, Peter. 2011. Fluxus Interruptus. Peter Wilkin's Blog. http://peterwilkin1.blogspot.com/2011/01/fluxus-interruptus-when-renga-joined.html. Accessed 24 March 2017.

CHAPTER 6

Evoking Narrative Landscapes with Mobile Media

In the previous chapter, I presented an account of how people use poetic forms to improvise and collaborate in mobile media ecologies. Now I turn my attention to place and landscapes in mobile media art practices. Again, I open with a vignette for your consideration.

An amateur local historian is having a meeting with an app developer. The historian has heard that it's pretty easy now to build apps. He has an idea for a walking trail that would take in architecturally significant houses of the mid 20th century. The local area is flush with prime examples. Sadly some have been demolished to make way for apartment blocks. The stories of how they were built were fascinating, he thought. There were artists and poets who'd lived here as well. Their poetry and art should be included in the app as well. He'd also like the app to show people how the streets used to look before the developers moved in. He wants it to be like a virtual guided tour that works like Pokemon Go. His granddaughter had shown him how it works. They went on a really long walk and it seems they used places of local interest for geolocated markers where you'd collect tokens to help you catch Pokemon creatures as well as earn points. You can see a map the streets around you. Apparently they'd used Google maps and GPS to do that. He liked the way you could see the landscape around the Pokemon you were trying to through the smartphone screen through the camera. This got him thinking about the app he envisaged. He'd like people to be able to add their own responses to the walking trail. These responses would be in the form of short poems like haiku or just short phrases. The poems and phrases could be layered over the significant

© The Author(s) 2017
M. Berry, *Creating with Mobile Media*,
DOI 10.1007/978-3-319-65316-7_6

sites so that people would see them if they hold the camera up to markers along the trail. He thought he might apply for some arts based funding so he could commission the project.

I begin this chapter with two key questions: How we can make poetic landscapes with mobile media? What are the practices around locative media and mobile storytelling that we can draw on to make poetic landscapes? Our local historian's design vision is complex. It is motivated by a desire to inspire a correspondence or dialogic interaction with the human and non-human elements of place and thereby enrich people's experiences places of local significance along a walking trail.

Before I introduce three creative writing projects that make poetic landscapes, I would like to provide brief definitions of two key terms of relevance here—locative media and mobile storytelling. Locative media art and mobile storytelling projects are reflective of our desire to imbue specific places with significance. Locative media is a term used to describe a form of mobile media that is emplaced in a specific physical site, accessible with a mobile computing device and uses a telecommunications system such as a mobile phone network. The term was coined at a media art workshop in Latvia in 2002 (Tuters and Varnelis 2006). It is a dynamic aspect of geographic places and may be ephemeral or a durable feature of the landscape.

In my previous work on locative media (Berry 2008), I observed that locative media art has some resonances with the Situationist movement of the 1960s where artists intervened in the urban landscape to provide alternate visions and readings of urban spaces. Where the Situationists used physical space, locative media artists use telecommunications networks as well to create contemporary commentaries of urban spaces. The spirit of the Situationist movement has been reincarnated in much recent locative media art. Random encounters in public spaces underpin the concept of psychogeography as theorized by Ivan Chtcheglov (1953). Psychogeography is a way of knowing the urbanscape through experience, for instance, walking. While walking, a person may have a chance encounter with something surprising. Such encounters are a common trope in many recent locative media works. McCullough (2006) argues that the application of locative media can contribute to enhance the richness of the city and identifies "urban markup" as something that can "attach as new layers to the forms and flows of the city". Urban markup is also a theme taken up by Farman who notes that it "can be done through durable inscription (like words carved into a durable façade of a building or statue) or through ephemeral inscriptions (ranging from banners and billboards to graffiti and stickers)" (Farman 2014, 4). Locative media art and mobile storytelling are ways to create maps and interventions in urban spaces that have multiple entrances and exits and can resist the forces that seek to exercise control through surveillance and commodification of public spaces. They are ways to set up correspondences and conversations between the human and non-human elements that may be found in places. Often they are instances what Farman (2014, 4) would call creative misuse of technology that point to emergent practices or transform views of the place technology has in society and in everyday activities.

Mobile storytelling is premised on "the idea that there is value in standing at the site where an event took place; far more than simply reading about an event, being in the place where that event happened offers an experiential value that gives us a deeper sense of the story and the ways that story affects the meaning of the place" (Farman 2014, 7). Embedded in this idea, too, is proposition that "mobile media offer the possibility to layer multiple—even conflicting—stories onto a single space" (Farman 2014, 8) in a way that is simply not possible in non-digital forms. Locative media art and mobile storytelling may add layers to specific GPS coordinates or may trace a person's trajectory or path through a place or landscape. The permutations are complex so to gain a clearer understanding of how locative, mobile and social media may intersect, I turn now to Tuters and Vangelis (2006) who classified locative media into two types of mapping—annotative and phenomenological.

Annotative mapping refers to the practice of virtually tagging the world. The other type of mapping identified by Tuters and Vangelis (2006) is phenomenological which refers to the practice of tracing the movements and/or actions of a subject. They cite Esther Polak's "Real Time" project where pedestrians with GPS devices move around Amsterdam leaving traces of their everyday routines as an example of phenomenological mapping. But their distinction is questionable because locative media can be annotative and phenomenological at the same time. Perhaps rather than viewing annotative and phenomenological as discrete classifications or categories, it would be more useful to see them as characteristics or qualities which can both be present in a locative media art work (see Berry 2008 for a full discussion). Over the past decade, locative media has become increasingly sophisticated and is affecting our relationships to place in both spatial and temporal ways.

In this chapter, I present three practice-led locative media art and mobile storytelling projects that show how poetic landscapes may be created. The methods implemented in these projects have strong resonances with Ingold's description of non-representational ways of working where a correspondence arises between "things and happenings going on around us, but of answering to them with interventions, questions, and responses of our own" (Ingold 2015, vii). But before I present these projects, I discuss theories of landscapes, places and spaces because these are integral to locative media art and mobile storytelling.

Landscapes, Places, and Spaces

The concepts "landscape" and "place" are conceptualized by and migrate across several disciplines including anthropology, archaeology, architecture, fine arts, geography and literary studies with each discipline adding their own particular nuances. In an earlier essay where I was theorizing these three terms (Berry 2013), I observed that Meike Bal would describe them as travelling concepts because of their transdisciplinarity. According to her, the value of travelling concepts lies in their ambiguity and propensity to slip between disciplines so that "while groping to define, provisionally and partly, what a particular concept may mean, we gain insight into what it can do" (Bal 2009: 17). (Interestingly, both landscape and place words can be used as either nouns or verbs.) Rather than disambiguating these concepts, in this chapter, I embrace the richness, intersubjectivity and contradictions inherent in the words "landscape" and "place" and treat them as travelling concepts—and as nouns and verbs.

In his essay "Scapeland", Lyotard invokes a notion of landscape that is rather visceral as well as visual where landscape is a "vanishing of a standpoint" through "an excess of presence" (Lyotard 1989: 216). Lyotard's landscape is a moment of complete presence, a state of mind where one is completely immersed in one's surroundings so that the idea of place ordered by knowledge recedes. Landscape is experiential and phenomenological when conceived of as a way of being interior to it. Landscape is also constructed through social practices and overlays whereby "environment manifests itself as landscape only when people create and experience space as a complex of places" (Knapp and Ashmore 1999: 21). We construct landscapes, inscribing ourselves into landscapes so that

> Landscape as practice or art practice is forwarded into process, as dynamic rather than either 'outside' experience or only focused through the physical character of encounters. (Crouch 2010: 14)

Being and walking in landscapes has an interior as well as an exterior dynamic so that a "landscape is situated in the expression and poetics of spacing: apprehended as constituted in a flirtatious mode: contingent, sensual, anxious, awkward" (Crouch 2010: 7), in short it is an experience that is mutable, complex and intangible, and it can inspire poiesis (Heidegger 1996). Landscapes are far from static. They are dynamic clusters of meanings and knowings that shift as we move through them our bodies, our senses, our thoughts, world views and feelings. How we attribute meaning to landscapes depends on many factors including our cultural capital—our habitus (Bourdieu 1984) as well as our lifeworld.

In a special issue of *Coolabah* focused on Australian geography and cultural production, Bill Boyd suggests that clear distinctions between concept words such as place and landscape be suspended in order that "the actual lived and experienced relationships between person and place/landscape" (Boyd and Norman 2013, 2) may be examined from different vantage points and disciplines. This represents a shift away from thinking which scientifically differentiates objects into categories but represents a more holistic logic that seeks to understand the complex systems of the everyday world in all their messiness from inside out rather than outside in according to a set of classifications or a taxonomy. A non-representational approach to understanding landscape can include associated sociocultural, historical and narrative practices as well as embodied multi-sensory experiences rather than through discrete categories. Furthermore, a geographic landscape is experienced as a complex of places (Knapp and Ashmore 1999). I would add that these also include the affective, imagined, projected and superimposed as well as the material and embodied.

Augé (1995) sought to account for the influence point of view has on perceptions of place in his conceptualization of place but his binary opposition between anthropological places and non-places is problematic because it implies that places are either one thing or another because of their intrinsic as well as functional characteristics. In his dichotomous model, non-places are associated with transit. Airports and shopping malls are examples of non-places. On the other hand, anthropological places are configured through inscriptions of identity, local references and knowledge of tacit rules governing interaction. However, places such as airports, shopping malls and railway stations also can be configured through inscriptions of identity and belonging, for example airports are workplaces for the staff who work there even though these are places of transition for travellers. Augé's binary model does not really take this into account. Hence, to avoid the pitfalls of classifying places into binary abstractions, I conceptualize all places as anthropological in this chapter.

The projects discussed in the ensuing sections involve community and/or crowdsourced mapping of places with narratives in the form of poems that will help us understand how places are inscribed with narratives of identity, local references and rules governing social interactions through a methodology informed by ethnography as well as creative practice research. The mapping and spatialization processes of making locative media poetry maps can assist our understanding of how people's perceptions of places play out across the dimensions of cultural, human, physical and spiritual geographies. Additionally, the aggregations of the poems pinned to place in the words of Farman are "multivoiced, layered, situated, and tell important (and often contradictory) narratives about a place and what it means to live in that place" (Farman 2014, 15). So, what does being in landscapes mean and what role can poetic maps play in drawing attention to the landscape? To address these questions, I present three projects that are part of a larger body of work that commenced over a decade ago. In this chapter, I consolidate and draw together research I have led, which has been presented in earlier essays and articles (see Berry 2008; Berry 2013; Berry and Hamilton 2010; Berry et al. 2011; Berry and Goodwin 2012) exploring relations between place and mobile technologies.

The Bigger Picture: Towards an Annotated Poetry Map

… stories we can walk into bodily,

stories we trace with our feet as well as our eyes.

(Solnit 2002: 71)

The *Poetry 4 U*: Pinning *Poetry to Place* project explores the implications of Web 2.0, smartphones and location-based technologies such as like GPS, Google maps for creative writing and arts practices. Like many creative projects, *Poetry 4 U* started with an intuition—an impulse to add a poetic layer to a landscape using location-based media technologies. The conceptual frame underpinning *Poetry 4 U* had its genesis in 2008. I had brought together a transdisciplinary team in order to experiment with locative media installations for artists and poets. Our first experiments were with a Bluetooth server with a GPS reading system.

Using a GPS reader, we staged three successful locative media installations of the Bluetooth server in 2009. One event included the delivery of curated poetry via twitter and the Bluetooth server as part of Mobile Textualism, Melbourne Writers' Festival, 21–30 August. We noticed that there was a massive shift towards iPhones in 2009 and a surge in iPhone application development. Bluetooth technology as a way of sharing information seemed to be a short-lived phenomenon within the context of mobile media. Advantages and disadvantages of Bluetooth and iPhone applications were each analysed and compared.

Bluetooth was found to be at a disadvantage when compared to simply using an Internet connection and the Google map API. Cost-wise, Bluetooth requires specialized equipment such as hardware installations for the Bluetooth relays. The Bluetooth platform also requires more involvement from the user to establish and manage connections, and this may prove to be frustrating with the short range of the Bluetooth network. User experience thus may be denigrated. At the same time, most smartphones are constantly connected to the Internet via mobile networks, which are widely available especially in an urban area. In fact, because location-based applications heavily rely on mobility, reliability and high availability, restricting use to Bluetooth spots may seem counterproductive and highly constrains the expansion or addition of future features. We turned our efforts to smartphones the curation of poetic landscapes using urban markup and annotated maps. We thought about ways we could use mobile media technologies to intervene in landscapes to disseminate poetic forms of content.

The Poetry 4 U website engaged with a wider international trend with regard to publishing and disseminating literary works. There has been an abundance of devices and platforms that have come onto the market including iPads, iPhones, Android phones, Kindle, all of which disrupted established traditional publishing models, pushing them towards multi-platform models.

Parallel to this had been a growth in small independent publishers. Furthermore, Web 2.0 technologies have resulted in an explosion of blogging and various interest communities. For example, there has been an exponential growth in poetry community websites such as dVersePoets and online literary journals as a form of publication. Some of these websites such as American Tanka, RiverLit and Frog Croon offer peer-reviewed or curated spaces. Indeed, there are even sites that aggregate what they consider to be the best online literary magazines and journals, for example the Every Writers Resource website complies a list of the best fifty on a regular basis. These developments of curated spaces potentially could be theorized as virtual communities (Rheingold 2000) and networks of creative writers that are an extension of the second wave of electronic media (Jankowski 2006), which build social capital based on "interpersonal trust, social norms and association membership (Jankowski 2006: 64).

The Poetry 4 U site was initially conceptualized as a curated space to connect with established communities and networks of practice that provide "the overall conditions and basis for interpreting and making sense of activities and events" (Jankowski 2006: 64). Clearly, the constellation of Web 2.0, smartphones and location-based technologies has had huge implications for creative arts practices. Established communities of creative writers have adopted them in various configurations to create new media forms as well as publication and dissemination systems. They have also facilitated collaborative activities, which are global as well as local. I presented examples of collaborative practices in the previous chapter.

The design process for the Poetry 4 U website sought to address how to use mobile media initially to engage readers and creative writers with Melbourne as a City of Literature. We were also looking for a model that could be transferred to other locations relatively seamlessly. Through this process, the project team found a way of pinning creative artefacts in the form of Twitter length poems to a city landscape in a way that is accessible globally as well as locally.

DREAMING OF AR (AUGMENTED REALITY)

We had a vision of poetry text floating over buildings in the city that would be visible using a smartphone camera. The poems would be lined up with particular buildings, and we could use GPS coordinates to set up bounds for each poem. At various intervals, the poems would change. We wanted the user to be able to share each poem through social media such as Twitter or Facebook. We envisaged a second phase of app development where a user could contribute poems as well to designated locations.

We chose Swanston Street between Victoria St. and Federation Square in Melbourne's CBD because it forms an axis through the city. It is a busy street filled with shops, cafes and bars. Trams run regularly up and down. We used Google maps to pin the poetry to our route using GPS coordinates that would feed into the application. The Google map provided another way to access the poetry along the route (Berry and Goodwin 2013). In my blog entry documenting the process of pinning the selected poems to the map, I stated that

> I have started pinning the winning entries from the Poetry4U competition to Swanston St. At first I thought I would pin the poems with explicit references to specific spots. This seemed to be a straightforward thing but it did not turn out quite that way. Some poems 'spoke' to other poems. I wondered if this was anything to do with the Twitter stream that was there when the poets contributed their poems to the stream? So I thought about this. Can I recreate the poetry Twitter stream within the interaction of the iPhone app? This is my design and curatorial challenge now.

Once I had completed pinning the poems to places, I observed in a blog post:

> I have pinned the poems to Swanston St using Google maps. This increases our options with the application because it means that people not actually present in the street can read the poems. I have decided that this is appropriate in the spirit of universal access.

My colleague Goodwin (see Berry and Goodwin 2012) designed a distinctive visual identity for the application and the Poetry 4 U project. His design concept centred on a "P" for poetry or poem as well as place, which would also represent a pin. The material was ready for conversion to an iPhone application. Operating systems are often updated, and this means that the longevity of apps is affected because of the need for updates to comply with new operating systems. We realized that we would need to have ongoing services from a technical developer, and we wondered what would happen if we wanted to add new content—would we need the services of a developer? We re-evaluated our desire for an app and decided a Google map would suffice because we could still achieve our objectives of creating a curated publishing space for poetry that has a direct relationship to Melbourne as a city, and the poetry map was accessible to anyone with a smartphone (Berry and Goodwin 2012).

Our solution was simple. Our model could be transposed into other locations by crowdsourcing Twitter length poetry and using the "P" to pin poems to places on a Google map. We could also preserve the site specificity of locative media (Farman 2014, 3) so that those who were in situ in Swanston St. would have a poetic experience of place. In the words of Sherringham, the Poetry 4 U project had become "a frame, but nothing that comes to fill that frame can be said to complete or realize the project, which always remains open and unfinished. Yet within its framework a shift, essentially a shift of attention, takes place" (Sheringham 2006, 146–147). In the next section, I discuss how the Poetry 4 U frame was used to curate community material in a remote region in Australia.

An Ethnographic Tour: Choosing Special Places

In early 2012, I was invited by Pilbara Writers group in Karratha to make a poetry map for the Pilbara region when they saw the *Poetry 4 U* website[1] where poems are pinned to geographic locations. I visited the Pilbara 17–23 June 2012 to commence the poetry-mapping project with members of the Pilbara Writers group (Berry 2013). Settings are integral to creative writing practices, for as Krauth observes in his discussion of the importance of place to writing:

> We make places make meanings for us – we imbue them that way – yet some places we acknowledge to be already imbued with significance, and to many others, of course, we assign no significance at all. (Krauth 2003: n.p.)

The Pilbara region comprises 505,000 km^2 (ABS 2008). Wikipedia describes the Pilbara as: "a large thinly populated region in the north west of Western Australia known for its Aboriginal peoples, its stunning landscapes, the red earth and its vast mineral deposits". The Pilbara contains three distinct geographic formations—the immense coastal plain, mountain ranges inland and an arid desert stretching right to the Northern Territory border. Its topography is characterized by gabbro and granophyte boulder piles ranging in colour from deep orange to purple, and spinifex grasslands. There is a diversity of habitats including snappy gums and hakea, which support various fauna and flora (Burbridge et al. 2006).

My research agenda on this field trip was straightforward: I wanted to document the process of selecting sites for the poetry map as well as to "activate a space and time within which I might engage with and explore issues of landscape" (Wylie 2005: 234) in terms of relationships for marking up places using networked technology and cultural production. The ethnographic approach I had chosen for this field trip allowed me to be in close proximity to the writers in the context of their lives so that I could gain a better understanding of their social and cultural practices and how creative writing fitted within that. Ethnography is a way of studying the meaning of the behaviour and interaction amongst members of a culture for as Spradley (1979, 5) states, "the essential core of ethnography is this concern with the meaning of actions and events to the people we seek to understand". As a research methodology, it is reflexive, exploratory, messy and inductive whereby the researchers immerse themselves in the context they wish to study. It enables what Geertz famously (1973) termed "thick descriptions".

I chose a non-representational hodological research strategy, and my actual methods included walking as well as photography and video. Hodology refers to the study of paths and has its origins in Greek, "hodos", meaning way, road and journey. Argounova-Low suggests, "the hodological approach underlines the very flow of both roads and narratives" (2012: 195). I felt this strategy would be incredibly useful because it would let me explore the connections and flows between poetic expressions, places and landscapes. I felt walking along paths was an essential method on this field trip for as Lee and Ingold argue, "walking allows for an understanding of places being created by routes" (Lee and Ingold 2006: 68). Arguably walking together with a participant also builds empathy through an immersive experience in their lifeworld.

As a researcher, I felt needed to gain and to "invoke the complexities implied by an anthropological use of the phrase—'a sense of place'" (Pink 2007: 240). Pink (2007) suggests walking with video is a way of experiencing place through the sensorium as well as emplacing the senses in order to get an understanding what a place may mean. By walking with video in the form of my smartphone when writers took me to their favourite places, I was able to document the relational aesthetics of these places, of being there together, so that I could empathically share the writers' sense of landscape, which inspires their writing.

I was conscious that I was walking in a landscape where Indigenous stories are woven through the land. As I walked with my camera and guide, I recollected reading the account of a Navajo poet driving through his land where "rocks are songs dappled with ancient starlight, suns, wind, and rain" (Bitsui 2011: 28). Here too the rocks seemed imbued with so much history of human habitation—over 40,000 years. So many feet had trodden here before me. I walked beside my guide who defined herself as a desert girl. My guide was not indigenous but had a rich knowledge of the cultural significance and stories of the landscape because she had lived there for a long time. She told me about her recent hobby of birdwatching. She identified it as a way in which she could feel a sense of kinship with Pilbara landscapes.

As we stood together, she pointed out a whistling kite and asked me if that was the type of bird I had seen hovering in the sky that morning, just after dawn near her place. It was. She confided her ambition to capture each bird of the Pilbara in a haiku. Birds were an integral part of the landscape for her as well as a source of poetic inspiration. The sight of the whistling kite propelled us both right into the landscape, into its presence to exclusion of all other standpoints (Lyotard 1989). For me now, the Fortescue River is overlaid not only with the Yindjibarndi story of the Rainbow Serpent but also with a magical whistling kite hovering, scanning the ground below for a meal before flying towards the horizon and vanishing into the air. In that moment, the landscape of the Fortescue River became vertical and timeless for us (Tuan 1974).

We continued our journey to Millstream Pool. Through walking together along the tracks, we were able to see place itself as a multi-sensory phenomenon, and I could empathically sense what was important to my guide (Feld and Basso 1996; Pink 2007). Sensory immersion in the landscape reminded me that it is never possible to describe an experience in its entirety (Husserl 1952/1932). We were both acutely aware that this landscape had aspects beyond our cognitive understanding. Paul Taçon (1999) has discussed the levels of sacredness layered over landscapes in Australia whereby as people gain different levels of knowledge through initiation, they receive access to sacred landscapes and sites. The paths along which we walking towards the pool with were "conduits along which narratives, memory and knowledge flow" (Argounova-Low 2012: 191–2). Crang suggests, "literary landscapes are best thought of as a combination of literature and landscape" (Crang 1998: 57). We were walking through storied or literary landscapes that have immense spiritual significance for Indigenous peoples in Australia. Many of the stories associated with the path to the Millstream Pool include the Rainbow Serpent (Juluwarlu 2007). The Yindjibarndi name for Millstream Pool is Chinderwarriner Pool. The pool itself is described as an increase or rainmaking site:

> This increase site is in Chinderwarriner Pool, near Millstream Homestead. The actual site is a termite mound submerged near the small island in the middle of the pool.

> Because it's underwater, you have to dive into the pool to work this site. To perform the ceremony that brings rain the *yurala* (rainmaker) puts their arms around the mound, takes some water from the pools in their mouth and spits it out again. They ask the site to bring rain, and tell it where and when they want it to fall.

> This site is one of two rainmaking sites, and is connected to the rain-making trees which are found near Chinderwarriner Pool. (Juluwarlu Aboriginal Corporation 2007: 38)

The other methods I used on this field trip to gather content for the Poetry 4 U project included semi-structured interviews, round-table conversations and observations. Participants were sourced through the Pilbara Writers' group. I also went on trips with volunteers to document the "special places" they wanted to include in the poetry map for the website. I kept notes and reflective diaries of my experiences and observations. I also led poetry workshops, which explored the writers' connections to desert landscapes. In the workshops, we explored how places in the Pilbara were imagined by the participants through conversations and the writing of short-form poems such as haiku.

I found that the poetry map offered the participants a new way of thinking about the geography of the Pilbara region. It offered them a path to accessing and sharing the often intangible dimensions of human geographies. The map, like the Swanston St. map, has been constructed using Web 2.0 technologies—it's an annotated Google map.

An aggregation of the poems pinned to the Pilbara region by the Pilbara Writers group points to broader narratives including the uneasy relations between the locals and the fly in fly out workers who work for the multinational resource companies. The social, economic and environmental impacts of mobility and transience emerge as key concerns of the writers. Another theme that emerges from the poetry map is the emotional entanglements people have with the dispossession of the Australian Aboriginal peoples and Land Rights struggles. Below is a haiku lament for lost knowledge pinned to the Burrup Peninsula on the Pilbara poetry map. It is by Delphine MacFarlane, a local writer:

Petroglyph power,
rock of ages history,
Jaburara dreams

The Burrup Peninsula is part of the Dampier Archipelago. The area has enormous national and international significance as cultural, archaeological and ecological site. McDonald and Veth (2009) report that there were over 1500 recorded sites in this region with petroglyphs, some of which are over 18,000 years old. Numerous motifs including anthropomorphs, non-figurative geometric shapes, animals, birds and animal tracks appear in the rock art. For Indigenous people, "these stories carved in stone are more than a past and a representation" (Morgan and Kwaymullina 2007: 10), and they believe that "destroying such places deprives us all not only of a past, but of a future" (Morgan and Kwaymullina 2007: 11). Sadly, the Burrup Peninsula is also the site of the Flying Foam Massacre in 1868 when the Yabujara people were systematically massacred and wiped out (Reynolds 1987). It is also the site of Woodside Petroleum's operations.

Taken together, the poetic expressions and stories contained within the Pilbara Poetry map provide unique and inflected points of view as well as the lifeworld of people who live and work there and who are interested in creating and sharing evocative expressions about their lived experiences. The poems pinned to the map are a window through which sensory, affective and spiritual states may be seen. There are poems about specific landscape features, sensory responses, and spiritual and emotional concerns connected to life in the Pilbara region. The Poetry 4 U frame has provided a way to bring these together to see the narratives that can emerge through poetry pinned to places. In the next section, I discuss the process of making the third Poetry 4 U map for the Geelong Writers Group. Geelong is a regional port city on Corio Bay about an hour's drive from Melbourne. Its population was estimated as 191, 440, June 2016, by the Australian Bureau of Statistics.

Bollards by the Waterfront

It's a windy day on the waterfront. Seven people from a writers group huddle over coffee. They chatted about the richness of the local history attached to the Baywalk Bollards walking trail. They felt it would be perfect for a poetry map. This would make a great collaborative project for the writers group as well. They envisaged the poetry map of the Baywalk Bollards walking trail as an anthology that would showcase local writers. The map was put together over series of writing workshops conducted between January 2012 and December 2013. The map turned the actual practice of creative writing into a social event for the participants. Below is a selection of poems pinned to the bollard locations. The bollards themselves are a form of urban markup that evoke historical people and events, and now, they have overlays contributed by local writers. The map is not yet complete and is open to further contributions. The bollards, which are the work of artist Jan Mitchell, are numbered and are integral to the identity of the Geelong waterfront. They depict characters from Geelong's history and are promoted as a tourist attraction. The bollards have added a rich version of Geelong's history and historically significant characters to the waterfront landscape.

For the writers' group, these bollards were a ready-made narrative ready for a poetic intervention through a mobile media overlay. Levine (2014) observes that numerous "mobile and locative media projects testify to the power of technology to introduce narrative as a way to augment, annotate, or add historical richness to public spaces" (144). In this project, the writers built on what was already there in terms of a walking trail to add a local grassroots identity to the landscape.

32: Salvation Army Woman

The Salvation Army has been active in Geelong since 1883.

> *Pyramid ... Salvos*
> *Phoenix rising*
> *Geelong breathes*
> *- Sandra A Jobling*

30: Carrie Moore
description
Born in Geelong in 1882, Carrie became a musical comedy star, performing in Melbourne at age 14 and in London age 21 and was best known for her role as the Merry Widow.

> *Lady in pink ... bygone*
> *New breezes wash*
> *Geelong's sea shore*
> *- Sandra A Jobling*

> *Chiffon light creeping*
> *Cheeky bunny*
> *Peeping*
> *- Sandra A Jobling*

29: Western Beach Sea Bathing Co.
description
Western Beach Sea Bathing Company swimmers (established in 1872). These are also portraits of politicians Gordon Scholes, Nipper Trezise, John Howard, Jeff Kennett and Sir Hubert Opperman as a young man.

> *Three dolphins pretended to be sharks*
> *And everyone scampered out.*
> *How we laughed at their joke.*
> *- E. Reilly*

28: Victoria Baths Swimmers
description
Swimmers characterising the Victoria Bathing Establishment (1870).

Mermaids in turquoise waters,
A white sun laves the sands,
Our smiles will melt salt.
- E. Reilly13: Bathing Beauties

description
The beach front was the venue for beauty competitions from the 1930's.

Balmy breezes
Caress them lovingly.
Trembling palm trees sway and rustle
Whispering anxiously to each other,
As the trio of bathing beauties nod in agreement.
Corio Bay sighs its fond farewell
Beneath their bare feet.
- Yura Reilly

CONCLUDING REMARKS

So, how can creative practice combined with ethnography contribute new knowledge to the field of human geography and vice versa? What new knowledge emerges from this approach? The evidence from the poems contributed to the Poetry 4 U website suggests that our understandings of place are temporal as well as spatial. By invoking the imagination through the use of poetic devices, spaces can be created to increase the public's engagement where perceptions and relationships to geographic environments may be altered. The evidence I have gathered so far in this research suggests that poetic expressions of place and landscape, particularly those that invoke landscape as a complex and dynamic system that is physical, cultural and spiritual, encourage new ways of understanding the places we inhabit. The new knowledge that emerges from this approach is that creative practices and associated cultural production including poetry, photography and video art uncover intangible connections to landscape that often remain outside of scientific evidence-based research paradigms.

I have presented a longitudinal and multi-sited participatory art project where crowdsourced poetry is curated and pinned using Google maps to specific geographic locations. It provides a prism through which to view social and mythical dimensions of emerging media technologies (Berry and Goodwin 2013). The Pilbara poetry map is a step along the path of the broader ongoing Poetry 4 U project towards understanding how social technology informs and contributes to symbolic understandings of places through poetic and narrative expressions. The Geelong annotated map pinning poetry to a walking trail contributes to local history narratives.

Mobile and locative media can show how people spatialize and attribute meanings to places they encounter and/or inhabit. In this chapter, through practice-led research combined with ethnography, I exposed some of the spatializing processes of creative writers exploring their favourite landscapes through their craft in order to better understand how they make sense of places, and how they make sense of themselves in various landscapes. The broader research project Poetry 4 U contributes to existing knowledge about creative practice, mobile media and human geography. The latest project on the website is a poetry map of Dublin which is supported by Poetry Ireland. I end this chapter with a poem pinned to this latest map from Poetry 4 U:

Trinity College - Bridget Walsh
Under the tree at Trinity back gate,
I'd see you waiting,
smiling as our eyes meet.
27 years, still our tree.

By Bridget Walsh

www.twitter.com/bridget_ie

NOTE

1. http://poetry4U.org.

REFERENCES

Argounova-Low, T. 2012. Narrating the Road. *Landscape Research* 37 (2): 191–206.

Auge, M. 1995. *Non-places: An Introduction to an Anthropology of Super Modernity.* London: Verso.

Bal, M. 2009. Working with Concepts. *European Journal of English Studies* 13 (1): 13–23.

Berry, M. 2008. 'Locative Media: Geoplaced Tactics of Resistance'. *International Journal of Performing Arts and Digital Media* Intellect, Great Britain 4 (2, 3): 101–116. ISSN:1479–4713.

Berry, M. 2013. Being There: Poetic Landscapes. *Coolbah: Special Issue* vol. 11: 85–96. ISSN:1988-5946.

Berry, M., and Hamilton, M. 2010. Mobile Computing Applications: Bluetooth for Local Voices. *Journal of Urban Technology* 17 (2): 37–55. ISSN:1063-0732.

Berry, M., and O. Goodwin. 2012. Poetry 4 U: Pinning Poems Under/Over/Through the Streets. *New Media and Society* 15 (6): 909–929. ISSN:1461-4448.

Berry, M., M. Hamilton, and D. Keep. 2011. Transmesh: A Locative Media System. *Leonardo* 44 (2): 162–163. ISSN:1530-9282.

Bitsui, S. 2011. Converging Wor[l]ds: Nizhoni Bridges and Southwest Native Communities. In *Blueprints: Bring Poetry into Communities*, ed. K. Coles. Salt Lake City: University of Utah Press with Poetry Foundation.

Bourdieu, P. 1984. *Distinction: A Social Critique of the Judgment of Taste.* Cambridge, MA: Harvard University Press.

Boyd, B and R. Norman, 2013. "Introduction to Coolabah special issue on placescape, placemaking, placemarking, placedness ... geography and cultural production" *Coolabah*, 11. http://revistes.ub.edu/index.php/coolabah/issue/view/1330.

Burbridge, A., N. McKenzie, S. van Leeuwen, L. Gibson, P. Doughty, and N. Guthrie, et al. 2006. Between Rock and a Hard Place: Rich Biological Patterns Amongst Ancient Red Rocks. *Landscope* 21 (3): 13–19.

Chtcheglov, I. 1953. "Formulary for a New Urbanism". Full text at http://bopsecrets.org.

Crang, M. 1998. *Cultural Geography.* London: Routledge.

Crouch, D. 2010. Flirting with Space: Thinking Landscape Relationally. *Cultural Geographies* 17 (1): 5–18.

Farman, J. 2014. 'Site-Specifity, Pervasive Computing, and the Reading Interface' In *The Mobile Story: Narrative Practices with Locative Technology*, ed. Jason Farman. New York: Routledge.

Feld, S., and K. Basso (eds.). 1996. *Senses of Place*. Santa Fe: School of American Research Press.

Heidegger, M. 1996. *Being and Time*. Albany: State University of New York Press.

Husserl, E. 1952/1932. *Ideas: General Introduction to Pure Phenomenology*. London: Allen & Unwin.

Ingold, Tim. 2015. 'Foreword'. In *Non-representational Methodologies: Re-envisioning Research*, ed. Phillip Vannini, pp. vii–viii. New York and London: Routledge.

Jankowski N.W. 2006. Creating community with media: History, theories and scientific investigations. In *The Handbook of New Media* ed. Lierouw L.A and Livingston S. 55–74. London: SAGE.

Juluwarlu Aboriginal Corporation. 2007. *Ngurra Warndurala Buluyugayi: Exploring Yindjibarndi Country*. Roeburne: Juluwarlu Aboriginal Corporation.

Knapp, A.B., and W. Ashmore. 1999. Archeology of Landscape: Constructed, Conceptualized, Ideational. In *Archeology of Landscape: Contemporary Perspectives*, ed. Ashmore W and Knapp A.B. London: Blackwell.

Krauth, N. 2003. Four Writers and their Settings. *Text Journal* 7 (1). http://www.textjournal.com.au/oct03/krauth.htm.

Lee, J., and T. Ingold. 2006. Fieldwork on Foot: Perceiving, Routing, Socializing. In *Locating the Field: Space, Place and Context in Anthropology*, ed. Coleman S and Collins P. Oxford: Berg.

Levine, Paula. 2014. On Common Ground. In *The Mobile Story: Narrative Practices with Locative Technologies*, ed. Jason Farman. New York: Routledge.

Lyotard, J.-F. 1989. *The Lyotard Reader*. Oxford: Basil Blackwell.

McCullough, M. 2006. On Urban Markup: Frames of Reference in Location Models for Participatory Urbanism, *LEA* 14 (3) Locative Media Special, http://leoalmanac.org/journal/vol_14/lea_v14_n03-04/mmccullough.asp.

McDonald, J., and P. Veth. 2009. Dampier Archipelago Petroglyphs: Archaeology, Scientific Values and National Heritage Listing. *Archaeol, Oceania* 44 (Supplement): 49–69.

Mining Jobs No Experience. 2012. www.Westjobs.Com.Au. The West Online Group Pty Ltd. Accessed 17 July 2012.

Morgan, S., and A. Kwaymullina. 2007. Solid Rock, Sacred Ground: Cultural Vandalism in the Pilbara. *Australian Feminist Law Journal* 26 (1): 9–16.

Pink, S. 2007. Walking with Video. *Visual Studies* 22 (3): 240–252.

Reynolds, R. 1987. The Indenoona Contact Site: A Preliminary Report of an Engraving Site in the Pilbara Region of Western Australia. *Australian Archaeology* 25: 80–87.

Rheingold, H. 2000. *The virtual community: homesteading on the electronic frontier.* Cambridge, Mass: MIT Press.

Sheringham, M. 2006. *Everyday Life: Theories and Practices from Surrealism to the Present.* Oxford: Oxford University Press.

Solnit, R. 2002. *Wanderlust: A History of Walking.* London: Verso.

Spradley, James P. 1979. *The Ethnographic Interview.* Belmont, USA: Wadsworth.

Taçon, P. 1999. Identifying Ancient Sacred Landscapes in Australia: From Physical to Social. In *Archeology of Landscape: Contemporary Perspectives*, ed. Ashmore W and Knapp A.B. London: Blackwell.

Tuan, Y.F. 1974. *Topophilia: A Study of Environmental Perception, Attitudes and Values.* New York: Columbia Press.

Tuters, M., and K. Vangelis. 2006. Beyond locative media. In *Networked publics*, University of Southern California. Available at: http://networkedpublics. org/locative_media/beyond_locative_media. Accessed 20 November 2010.

Western Australia Planning Commission and Department of Planning, Government of Australia. *Karratha: Regional Hot Spots Land Supply Update.* Dec 2010. Accessed 17 July 2012.

Wylie, J. 2005. A Single Day's Walking: Narrating Self and Landscape on the South West Coast Path. *Transactions of the Institute of British Geographers* 30 (2): 234–247.

CHAPTER 7

Making Films and Video Art
with Smartphones

INTRODUCTION

Over the past decade, mobile media has changed our social, emotional and social cartographies and this has provided new opportunities for creative practice. I have been researching various aspects of mobile media through creative practice and ethnography since 2004, when mobile phone cameras were seen many practitioners as a mere toy, not to be taken seriously. Since then much has changed. The technological advances in mobile phone technology have been midwives for new spaces where serious film-makers can play.

Openings for image making and sharing (Schleser et al. 2013; Keep and Berry 2014) are now constantly embedded in the background of our routines because of our smartphones. This has changed the scene for film-makers and photographers in subtle and not so subtle ways. New creative vernaculars, both verbal and visual, are constantly materializing as established in Chap. 2. Our ability to play with smartphone camera apps and share our moving images with location-based overlays is giving rise to new opportunities. Smartphone cameras can no longer be readily dismissed as a toy, rather they are better considered as vehicles for serious play and creative practice.

The confluence of smartphones and mobile media certainly do remediate old media forms (Bolter and Grushin 1999) but they also shape storytelling to generate innovative forms engaging the imagination in novel ways. I will discuss some key examples of film-making in mobile

© The Author(s) 2017
M. Berry, *Creating with Mobile Media*,
DOI 10.1007/978-3-319-65316-7_7

131

media contexts including networked collaborative and transmedia film-making, digital storytelling and participatory video art and documentary projects to show how the field of film-making has been expanded and reimagined through mobile and social media. The kinds of relationships mobile film-making has with contemporary everyday life with its affective and sensory dimensions will be mapped through these examples and placed within a broader context of emerging visual vernaculars that is informed by my own current digital ethnography into the connections between mobile and social media, and creative practice.

In a previous account of mobile film-making (Berry 2017), I suggested that it is fruitful to situate mobile film-making through three concepts drawn from digital ethnography, namely wayfaring, co-presence and mobility placed within mobile media ecologies to see how emerging everyday creative practices and evolving aesthetics manifest. Here, I ask—how we better understand practices associated with making films and video with smartphone? In this chapter, I build on some of the material presented in my earlier essay (Berry 2017) to examine the implications of smartphone assemblages for film-making and explore new forms, contexts and agencies through the lens of non-representational theory to seek an emergent ethos of liveliness and animation (Vannini 2015). This chapter is arranged into four parts—evolving practices, emerging forms, changing contexts and adapting agencies.

Evolving Practices

It's 2006, Waikiki Beach, morning of 15 October, about 7 am. I was there for a conference and was presenting my paper at 2 pm that day. My hotel was a standard 3 star—basic and clean. I felt I had been very fortunate to be given a corner room on the top floor—the views were gorgeous. I thought I would go for a long walk along the beach after breakfast before going to the conference venue. I sat on my bed putting on my shoes. The bed swayed and vibrated. I heard a rumbling. I tried to stand up. The floor was swaying drunkenly. I remembered that I was in a volcano zone and wondered if there'd been an eruption or was this an earthquake? I had no idea. In any case, being on the top floor of a twelve-storey hotel did not seem like a good idea. I went into the corridor to see if anyone knew what was going on. There were people out in the corridor—all Japanese and they had pillows and blankets—when they saw me they shouted "*djijin*". We all dropped to the floor as a very strong after shock hit. My room door had slammed shut, I felt in my pocket, no card. I was locked out. I went to get in the lift—they pulled me back shaking their heads. We climbed down the stairs and the power went out. As we descended in the dark we encountered more people. We felt another aftershock or was it just our legs shaking from stress? It was hard to distinguish.

There was a tsunami warning in place. I figured it would be safe enough to go the hotel where the conference was held. There were hotels and high rises lining he streets all the way. If I heard the siren, I'd just go to the nearest building—no one would turn me away. I set off. And I started filming with my Nokia camera phone as I walked. There were queues outside the convenience stores; people were talking on their mobiles to people on the mainland of the USA to reassure them that they were okay and to get the latest news about the earthquake situation in Hawaii. There was no local news, no information. The airport was closed. The power was out—much confusion and shock but no real chaos. Later, I'd use the footage to create a video artwork that I'd call 6.5 after the magnitude of the earthquake. At the time though, I felt I needed to document the experience. I filmed everything without really discriminating. It dawned on me as I walked along filming that I was working in a tradition that had its genealogy in the early twentieth century with people like the Russian film-maker Dziga Vertov. The early pioneers of movies filmed everything they could —the stuff of ordinary life that was already there.

In *Man with a Movie Camera*, Vertov captured the minutiae of everyday life. His film is an ode to modernism in a post-revolution Russia keen to reinvent itself as a progressive and successful nation, but it's the film's ability to reshape and frame the city via the use of the film camera and cinematic techniques that best informs my creative practice. In Vertov's search for Kino-Pravda ("Cinema-truth"), he adopts innovative film techniques such as hidden cameras, filming atop buildings and moving vehicles, resulting in a film that provides the viewer with unusual and innovative perspectives of a city and its inhabitants. Vertov used film to create a new language of cinema where the "kinoglaz" (cinema eye) captured everyday ephemera to create a new form of creative expression through montage. He magnifies everyday events to expose the mechanics of social activities, thereby to make the taken for granted mundance details become the centre of attention. Later, Keep and I were to conceptualize mobile phone cameras as "mobiglaz" and Keep developed his notion of a "liquid aesthetic" (Keep and Berry 2014).

I possessed a mobile phone that could take photographs and video sequences and was with me all the time. I could take photos or shoot video whenever the impulse arose. This had a substantial effect on my creative practice. I experimented with low-resolution images and video with my camera phone. My other noteworthy early mobile film-making project was earlier in 2006. For the Commonwealth Games 2006, in Melbourne Australia, Mick Douglas invited a team of transport decorators from New Karachi led by Nusrat Iqbal, to decorate a Melbourne tram with chamak patti which in Karachi is used to decorate W-11 minibuses. I shot video on my mobile which I composited into a grid to allow different perspectives on the screen. The outcome was *Liminal Tram*—a single channel video designed for large screen projection.

My video fitted within a broader innovative and emerging cultural practice to document experiences using camera phones to produce digitally mediated memories. *Liminal Tram* was shown as part of a group exhibition at the Karachi VM Art Gallery 7–22 September 2006. Other artists included: Wajid Ali, Mick Douglas and Karen Trist. It included a catalogue essay written by the curators, Mick Douglas and Durriya Kazi. The exhibition was reviewed in Karachi newspapers: e.g., Kolachi, 17 September 2006. By popular demand, the W11 Karachi to Melbourne tram returned to the City Circle route. Because of *Liminal Tram*, I was invited to be an official video artist for the duration of the project (November 2006–March 2007). The project may still be viewed on the official website: http://www.tramtactic.net/W-11/. Shalini Singh reviewed the return of the Melbourne to Karachi tram in the Bharat Times, November 2008, 14.

It was also the time when video shot on mobile phones was creeping into mainstream news broadcasts after the London bombings of 2005 when visual material from people's mobile phones was used by all the major news broadcasters. The low-resolution pixelated aesthetic provided the visual material shown with a sense of urgency and authenticity and placed the viewers at home in their lounge rooms within the frame rather than as detached observers. The cameras were obviously hand held and people were shooting whatever was before their eyes without consciously framing specific instances. Typically, footage was abrupt, shaky, filled with digital artefacts and visual glitches. The sound was raw. Often sound tracks were almost completely muted to make space for the reporter's voice. The news channels had caught on and used mobile phone footage and images to add a cachet of authenticity to their reports. Mobile phone video had arrived in mainstream media as a new form of moving images.

EMERGING FORMS

The coupling of film-making and mobile phones has had profound impacts in media ecologies. A relational approach to mobile film-making, which considers the significance of emergent social practices and visualities as well as the reverberations of changes to technology, offers different perspectives to think about opportunities for creative practice. In the introduction to this chapter, I presented accounts of two early creative practice research projects of my own situating them with a genealogy of experimental practices with movie cameras. In this section, my focus is upon how film-makers and artists are using smartphones to remediate old forms (Bolter and Grushin 1999) and to create new forms. I will do this through discussing examples drawn from the creative practice research of others and myself.

Dean Keep is a mobile media artist scholar who has been making short evocative films with smartphones and the earlier camera phones for the past decade. He has argued that in the decade 2004–2014, the camera phone has reconfigured our relationship to photography so that traditional tropes associated with film-making and photography have been reimagined and remediated, noting that the "Kodak moment" has been replaced by the "mobile moment". He suggests that camera phones can be thought of as portable visual diaries where narratives of everyday life may be collected and shared "promoting a 'liquid aesthetic' whereby changes in technology, as well as our relationship with personal computing, may be understood as key drivers which shape image making practices" (Keep 2014, 23). He proposes that the absence of established conventions "promotes a culture of creative experimentation and presents new ways to view our world" (Keep 2014, 19). The interplay of mobile technology and social practices native to camera phones is central to Keep's notion of a liquid aesthetic and his experimental video art pieces.

In 2007, Keep and I initiated an exhibition with two photography academics, Pauline Anastasiou and Karen Trist called "Order of Magnitude". The group show was held in Melbourne at a university gallery space and in Brisbane at the Queensland Centre of Photography. The show had its genesis as a response to our critical colleagues in digital media and photography who felt that we were wasting time researching the possibilities of mobile phone cameras and screens for art making. The group show was also an outcome from our creative practice research. We worked from the proposition that the advent of the mobile phone camera transformed image capture from a consciously planned activity to one that can occur spontaneously. In the past, the use of movie and still cameras necessitated a conscious act: unless people remembered to carry cameras, they could not take still and moving images of events, people and places and mobile technology changed this irrevocably. In 2007, if one carried a mobile phone with a camera, no longer did one need to wish that one had remembered to bring a camera. But the images were low resolution and were not taken seriously. Indeed, they were seen as inferior in the realm of digital photography aficionados and professionals because of the prevalent obsession with high-resolution images that were realistic, sharp and clear without digital artefacts.

The blurb in our publicity described the show as a photographic exhibition embracing the low resolution qualities to explore the potential of the mobile phone as an image-making device. Our research questions were: What happens when you take these little low-resolution images off the phone and blow them up into large photographic prints? What happens when you take low-resolution video files shot on a mobile phone and project them on large screens in gallery spaces?

Our videos were experiments exploring how we could leverage the kinds of images camera phones typically produced. We noticed through our creative practice that some images were quite similar to images made with plastic cameras such as Diana, Holga and Super 8 movie cameras. In images and video taken with mobile phone cameras, we found that light can create unexpected effects. Composition can be difficult, especially in bright light when the screen is difficult to see. Colours can be distorted. When blown up the images contained echoes of Impressionism, Pointillism or Fauvism because of digital artefacts. The videos often took on a nostalgic cast because of the apparent surface skin created when projected onto a large screen. We decided that these qualities were highly suitable to works that pursued notions of memory and remembering. In "Order of Magnitude" we presented works that engaged with feelings of vicariousness, belatedness and displacement in relation to actual and imaged past events. The feedback left in the comments book in the gallery space indicated that our experimentations with mobile phone images and video had succeeded in engaging with earlier art movements such as Impressionism and with fine art photography. Through our initial creative practice culminating in the "Order of Magnitude" exhibition we understood that mobile phones and their subsequent evolution into what we now refer to, as smartphones would have huge ramifications for film-making and photography.

To explore some of these ramifications and potentials, I now turn to the work of Leo Berkeley who was also an early experimenter with mobile media. His initial forays included making machinima to be viewed on mobile phone screens, however, what is pertinent here is his short film titled *57 Tram*. This is a self-referential essay film that underlines the idea of mobility, experience and presence. Through the creative practice of making this film, he investigates the capacity of smartphones as "portable and accessible motion picture technology" (Berkeley 2014, 27) to offer "a personal, reflective style of documentary film-making such as that modeled by the essay film" (Berkeley 2014, 27). He suggested that there were some similarities to amateur still photography in that his improvised shooting process on the number 57 tram in Melbourne was "impressionistic, opportunistic, and fragmented" (Berkeley 2014, 30). This extended his practice into new directions where he was able to film moments that captured his attention and yet this very ability left him with a new dilemma which was "how to structure the mass of fragmented material collected in a way that conveyed the complexities of the lived experience while still being sufficiently coherent to interest the proposed audience" (Berkeley 2014, 31).

Berkeley found that the experience of making the film itself provided him with ways to rethink forms of mobility as well as interrogate the implications of the extreme accessibility of portable and potentially invisible movie cameras in the form of camera phones. He uncovered some personal ethical and safety considerations when he reflected on his unwillingness to document "the rowdy men who were almost always at the back of the tram" (Berkeley 2014, 32). For him, the extreme accessibility of a camera did not create the sense that "it was possible to film anything at any time" but rather that he did not want to distort an experience by conforming to the traditional conventions of film-making surrounding the need to include dramatic action as a part of his narrative. The *57 Tram* has the distinct voice of a lyrical essay, there are no turning points; rather it is a sophisticated and nuanced unfolding of lived experience tracing the natural contours of the tram route and journey. The film is also an extension of his research into the possibilities of smartphones as a production device for film-makers with a limited budget and resources and is an eloquent demonstration of how smartphones take film-making into new poetic forms such as the lyrical essay film.

Themes of the local and the personal are also evident in Patrick Kelly's films and creative practice research. His doctoral study (2013) comprised a film and exegesis where he explored the relationship between film-making and technology through a hybrid methodology that drew on practice-led research and auto-ethnography. His film *Detour off the Superhighway* was a screen-based creative practice experiment documenting his thoughts and journey over 80 days as he gradually eliminated digital and analogue film-making and photography technology. His experiments with the early forms of image-making technology are eloquently captured in this film.

In an article exploring how smartphones may take film-making into new areas and forms, he proposes that the "development and rapid uptake of platforms and applications to capture and share video are generating emergent practices associated with social media, presenting new opportunities for filmmakers to explore different contexts" (Kelly 2014, n.p.). He continues to explore the theme of slow media and how media encroach on everyday life in his creative works. In his auto-ethnographic film *North*, Kelly acknowledges traditional film-making genres and forms and uses these as playful, and at times, ironic counterpoints to the new creative vernaculars associated with emplaced visuality and sociality. He used the mobile video-sharing platform Instagram to reflexively engage with his experiences while moving through Melbourne as a juxtaposition of moments. He suggests mobile filmmakers "might discover the emergence of even more contexts and auratic experiences" (Kelly 2014, 136) because of the inherent "juxtapositioning nature" (Kelly 2014, 136) of contemporary mobile media platforms.

The ways in which juxtapositions can happen in digital media platforms and storage systems are also really important to the film-maker and screenwriter, Bettina Frankham (2016). She recently remarked on this tendency of smartphones to juxtapose content in relation to her own creative practice with smartphones, which she describes as "cinécriture" following Agnés Varda by saying that the "phone is my notebook, my mood board, the memory folder where I gather video, sound, still images and jottings of ideas, all to be worked further at a later date" (2016, 50). Frankham's argument has an elegant simplicity that extends both writing and film-making into new forms. She conceptualizes forms of writing with digital screens like her smartphone whereby writing happens as she records images and shoots video rather than writing being something that happens away from the screen so that making videos can become a way of writing rather than the final stage or outcome of a scriptwriting process. For Frankham, the juxtapositions that happen through the search algorithms in data storage systems such as the cloud or they appear serendipitously in her content folders have become integral to her creative practice. She concludes her argument for the need to rethink the forms screenwriting may take so that it includes writing with digital screens and that "literacy in these new territories of creative practice enables a considered and deliberate approach to forms of interhuman and human/algorithm collaboration" (Frankham 2016, 64).

How smartphones have impacted on screenwriting and film-making is mapped out by Batty (2014) in his discussion of the kinds of apps that have been developed specifically for screenwriters. He refers to McKie's app called Scenepad which can "provide factual development and collaborative tools to aide the development of a screenplay, namely through the writing and weaving together of individual scenes" (Batty 2014, 110). Scenepad can supply useful information for script development. McKie took this idea of a screenwriting app for smartphones a step further in 2013 when he developed Scenetweet, inspired by Twitter and its constraint of 140-character length posts. The app offers "screenwriters the opportunity to write in short instalments, on-the-go" (111) and to get feedback from other writers. Batty concludes that apps add to a screenwriter's arsenal and can extend creative practice into new forms and frameworks but that a screenwriter's "existing knowledge and methods" (112) remain paramount. In other words, while the technology may take screenwriting into new forms, the technology is no substitute for craft and as a tool is only ever as good as the person wielding it.

I turn now from Frankham's (2016) and Batty's (2014) discussion of mobile technology and new forms of screenwriting to Adam Kossoff, an artist film-maker who suggests that the mobile phone "allows us to imagine that we are part of the flow of everyday life" and argues that mobile phones may now be seen as an "inherent part of the event, as theorized by Alain Badiou" (Kossoff 2014, 40) where the event is tied to a zeitgeist and some kind of unpredictable and major social, political or personal disruption. He sees mobile phones as double edged where they are both "an object of desire" (Kossoff 2014, 43) that can give us a type of agency in the world yet at the same time they are "a fetish that obscures us from the world" (Kossoff 2014, 43). In his film, *Moscow Diary* (2011), Kossoff revisits the streets and buildings in Moscow, which Benjamin mentions in his famous diary from his visit to Moscow in 1926–1927. The film shot on a mobile phone camera engages with the modernist notion of a flâneur as an "urban wanderer who individualizes time and space by remaining outside of the metropolitan crowds and the cut and thrust of the city" (Kossoff 2014: 39). This work has elements of evocative auto-ethnography as explicated by Ellis et al., as he entwines his lived experience with Benjamin's diary entries "to produce aesthetic and evocative thick descriptions of personal and interpersonal experience" (2010, n.p.). The three aspects of mobility play out through an intricate braided narrative to bring in a fourth aspect of collapsed time that both obscure and enhance his movements through contemporary Moscow.

Changing Contexts

Social media and Web 2.0 technologies have been with us for over a decade and have had a profound influence on film-making. To explore how these have manifested, I would like to commence my discussion in this section with a description of contemporary mobile media ecologies. Mobile media has pushed film-making and photography into new terrains where old ways of working are being profoundly challenged. In his review of a decade of mobile phone film-making, Schleser argues that the accessibility to film-making technology has clearly increased through the ubiquity of mobile phones and smartphones and that this, in turn, is "shaping not only new modes of film production but also modes of consumption, distribution, and exhibition, by embedding these digital stories in network media" (Schleser 2014, 155). Schleser concludes his review of mobile moving-image practice to claim that smartphones "provide prospects for the twenty-first century citizen to develop innovative and imaginative cultural competencies" (2014, 167).

The profusion and reach of smartphone camera applications mean that tools and editing techniques once only available to professionals are now readily accessible. We can all be mobile film-makers now, provided we have smartphones, video-editing applications and are prepared to play in mobile media ecologies. This has had a substantial impact on what it means to be a film-maker and on the art of film-making itself. Felipe Cardona observes that ready access to film-making technology has the potential to disrupt more traditional aesthetics:

> Today, when YouTube allows any user to upload and exhibit his or her videos to the world, there is the possibility that uninitiated users and producers of video material can dare to take montage to its limits... New technology and workflow democratisation allows end users, now prosumers, to teach television channels, film studios, and even mainstream record labels, the future course of music, film, and television. (Cardona 2014, n.p.)

We may think about film-making in and for mobile media ecologies simply from a perspective that focuses on representational dimensions. However, such perspective fails to account for the dynamic intricacies of mobile media ecologies and the ever-changing media practices they enable. And as such, what do the dynamics and intricacies of mobile media ecologies and smartphone cameras offer film-makers? To better understand both the potentials and challenges for playing creatively with smartphone video in complex media environments, we should consider the implications of mobility itself along with networked co-presence. Mobility as theorized by Cresswell (2010) operates within mobile media ecologies where practices and actions that may be described emplaced visualities (Pink and Hjorth) are becoming increasingly taken for granted.

Mobility has become a dominant theme in contemporary mobile film-making as film-makers grapple with the extreme accessibility of cameras and the potentials of playing with their practice in mobile media ecologies. Many of the films submitted to MINA (Mobile Innovation Network Aotearoa) over the past six years have been concerned with mobility and were shot on public transport or in cars. So what is it about mobility that so fascinates mobile film-makers? Why are so many of the entries into the MINA film festival shot out of moving cars and trains?

Mobility as defined by Cresswell (2010) can help address these questions to better understand film-making practices in contexts which are overlaid with mobile media and where access to cameras may be taken for granted. To elaborate further, in the contexts of mobile media ecologies, people are often moving through physical environments that are separated by timezones and geography yet at the same time they are connected with each other through networked technologies. Mobility provides film-makers with novel and playful ways to think about how the "parameters of the camera phone" (Berry and Schleser 2014, 5) can inspire new creative practices. Cresswell (2010) provides us with a way to conceptualise mobility and mobile environments that goes beyond representation. He proposes that we look at mobility as three elements or elements that are bound together—movement through the physical world, representations of movement, and practice as an embodied phenomenon.

The work of Gerda Cammaer is an eloquent example of how smartphones and iPads have shifted film-making into new contexts that engage with both the tropes and the corporeal dynamisms of mobility as conceived by Cresswell (2010). She has extended the form and contexts for travel video with her film *Mobilearte*. The blurb on her Vimeo profile describes the film thus:

> Mobilarte is based on a video recording of a tuktuk ride through the city of Maputo (Mozambique) filmed on an iPad. The piece explores the particular qualities of iPad moving images (ghost frames, pixellation), as well as various ways to represent quick impressions and fleeting memories of life in Maputo. The video is edited following the workings of associative memory and change blindness, the fact that if movement and action remain constant, viewers don't perceive continuity errors in film. The sound is constructed by creating four different musical tones of short tuktuk sounds, which were each attributed to one of the national colors of Mozambique: yellow, green, red and black. The actual sound track is the result of applying a computer program to the image that scanned it for those four colors and translated this into music. The video has three parts that each have a different emotional character: Maputo Alto, Maputo Baixa and Maputo Praia.[1]

Her oeuvre embraces mobility and pushes the boundaries of film-making through poetic juxtapositions of visual and sound images. She exploits the characteristics of video shot on mobile devices to emphasize aspects of mobility. A more recent work, *Kinetic Travel Memories #1*, also shot with an iPad plays with the new contexts of mobile media further and is described by Cammear as "an experiential representation of the experience watching both the view out of the window and the iPad's rendering of that view at once. The video is meant to evoke the sensation of speed and (cosy) displacement when travelling (in this case by train)".[2]

The ability to experiment with smartphone camera apps and share any resulting moving images through social media with location-based overlays has given rise to new ways of communicating visually using smartphones and other networked mobile devices in our everyday social interactions. In a conversation about his film *Give Me Faces Give Me Streets*,[3] Kelly told me that his ideas for film-making were inspired by the idea of enacting and re-presenting emplaced visualities. The work is auto-ethnographic in that it uses lived experiences reflexively to engage with well-trodden path of the dichotomy between the urban and the rural. This is a continuation of the creative practice methodology he devised for his Ph.D. and which draws on evocative auto-ethnography as conceptualized by Ellis and Bochner (2006). The project itself arose out of collaboration between Kelly and another film-maker Kelly Shanahan where she challenged him to explore the urban/rural dichotomy through sound. He used his iPhone voice memo app to record sounds in the city and in the country. The country sounds include the crunch of footsteps on gravel and wind distortions. The city sounds include voices, traffic, and building construction. He then overlaid an audio of himself responding to Walt Whitman's poem "Give me the splendid silent sun". Kelly recorded the visual material using Google Street View within the Google app on his smartphone. The work extends both the forms of mobile film-making and has created a new context for creative practices with smartphones. The film was selected through a double peer-review process for MINA, which showcased mobile film-making projects from around the world in Melbourne (Lido Cinemas, 1 December 2016) and Wellington (Ngā Taonga Sound & Vision, Te Anakura Whitiāhua, 1 December 2016).

Keep's lyrical video artwork, *Decombres* is shot out of a train window on the route between Melbourne and Ballarat. When he reflected on the piece in a conversation, he told me that the voice over was a series of texts between himself and a friend and he had them translated into French to add a pensive atmosphere as well as to evoke other train journeys he had taken while in France.

ADAPTING AGENCIES AND AMBIGUITIES

Mobile media generate spaces where agencies are constantly evolving and adapting and where ambiguities thrive. Recently in *The Mobile Audience*, Martin Rieser (2011) asked, "in what ways can the new modes of audience engagement and participation in dispersed or interactive artworks be evaluated?" This is a complex question that would make an excellent thesis topic. My aim in referring to his question here is not to provide answers but rather to use it as a touchstone for my discussion of mobile media and agency in this section.

These shifting dynamics of agency, audience and participation provide rich conditions of possibility for film-makers and other creative practitioners who work with video and moving images. The ability to tether mobile and social media together has led to the emergence of new visual vernaculars—which I have discussed in depth in Chap. 2—as well as networked socialities and what Pink and Hjorth (2012) conceptualize as emplaced visualities. They define emplaced visuality in connection with camera phone practices to include tensions between representational and non-representational aspects of smartphone photography in dynamic and complex mobile media ecologies. The boundaries between physical and social media environments become blurred to the extent they become obsolete or superseded because of the ability to move seamlessly between the physical and online spaces. Remediated (Bolter and Grushin, 1999) hybrid and new forms flourish in video sharing and locative media sites.

The affordances of smartphones allow anyone to make, edit, geotag and share photographs, video works and films, in other words, smartphones place the means of production at the disposal of many. In the introductory chapter to *The Mobile Story*, Farman (2014) notes that mobile locative media disrupt the ways in which we think about the production and distribution of narratives because "emerging mobile stories are multi-voiced, layered, situated and tell important (and often contradictory) narratives about a place and what it means to live in that space" (Farman 2014, 15).

However access to the means of production for such multi-voiced narratives as described by Farman (2014) does not necessarily translate into access to distribution channels. Ethnographer Patricia Lange argues, "the basic premise of interactivity in online, video-sharing sites interrupts a commonly accepted notion of the centrality of the video maker as a strictly independent force behind particular content" (2017, 145). She points out that "participants, audiences and scholars often see a distribution site as merely a neutral platform or final step in the act of video creation and sharing" rather than as entities motivated by financial imperatives and "even vernacular video makers who are not necessarily interested in making money or becoming famous are nevertheless ensconced in the commercial infrastructures and rules that undergird distribution sites" (2017, 149).

On the other hand, even though commercial infrastructures continue to dominate distribution, participatory media can give expression to oft-marginalized voices. Jennifer Deger, anthropologist and curator works on practice-led digital media projects with Aboriginal people in north Australia where they "use photography, video and installation as a means to renew relationships between kin and country, giving life to ancestral connections while simultaneously generating new relationships with people and places far from home" (Deger 2017, 319). Smartphones were integral to the process of making media for the installations and also set up relations between curated media on display and visitors to the exhibition. These assemblages give agency and compress geographic distance for such communities despite the ambiguities of access to mobile media.

We are immersed in a liquid modernity where geographic "distance is no obstacle to getting in touch, but getting in touch is no obstacle to staying apart" (Bauman 2003, 62). While Bauman claims that the "advent of virtual proximity renders human connections simultaneously more frequent and more shallow, more intense and more brief" (Bauman 2003, 61), I would strongly suggest that it is this very paradox that makes social media so rich in opportunities for creative practice. It is easy to join and to leave collaborative projects that occur within mobile media ecologies.

The participatory mobile media film projects I present in rest of this section show locative media overlays combined with physical places to create polyphonic narrative collages that traverse both time zones and geography. The creative premise or theme of gathering together filmed expressions of place and mobility over a 24-hour time frame across global time zones using location-based technology is emerging as a dominant and common trope in contemporary mobile film-making.

The first project I wish to discuss is part of Iceberg Fernandez's Ph.D. practice-based project at the University College of the Arts, London. It is called *NOW&HERE = EVERYWHERE* and is an excellent example of how research can be disseminated and integrated into the everyday life of people other than the candidate or researcher. The project illustrates how social media participatory cultures serve as a source of inspiration for creative practice research as well as a way to socialize through emplaced visuality. The text below is an extract from my Facebook timeline in May 2013:

> NOW&HERE = EVERYWHERE is a Quantum Filmmaking project, participatory mobile phone cinema, celebrating cultural diversity, in which you can collaborate with your mobile phone.[4]

It is an excellent example of how research can be disseminated and integrated into the everyday life of people other than the candidate or researcher to create and claim an impactful pathway and outcome for the research. We can see Bauman's paradox of virtual proximity at play, and in this case exploited by an artist to create a new work through emplaced visuality. Participation is low risk in the sense that it can end "without leftovers and lasting sediments" because it relies on a virtual proximity, which can be "both substantively and metaphorically, be ended with nothing more than the push of a button" (Bauman 2003, 62). While Bauman laments virtual proximity as being a cause of weakening social bonds, within the context of participatory art projects it provides a condition to enable greater access for those who would like to become involved with global locative media projects to express a point of view within predefined parameters.

The second project I present was curated by Max Schleser and it has strong resonances with Fernandez's *NOW&HERE = EVERYWHERE* project. Schleser's *24 Frames 24 Hours* project (see Schleser and Turnbridge 2013) aimed to capture life and a perception of place in different cities over a 24-hour period and has developed into a dynamic website[5] where visitors are encouraged to contribute their work and are provided with tutorial so they can access clips from parts of the world that interest them to encourage further participation. The project was collaboration between students at Te Papa Museum, Massey University, New Zealand and the University of Padeborn in Germany. Forty-three film-makers working only with mobile phone cameras created twenty- three two-minute videos. Online video conferences were used to workshop ideas and provide feedback. The videos produced were pinned to a Google map so that viewers can see exactly where they were shot. Location based technology provided the film producers with an innovative way to present their work that moves away from more traditional linear show reels and where the viewers may order the films according to geographic location. In this way, the project invites viewers to become digital wayfarers (Hjorth and Pink 2014) who can trace paths to create their own journeys through *24 Frames 24 Hours*.

The third project I present also illustrates participatory locative media video and mobile film-making. The project is Vickers's *24-hour.in*[6] which was inspired and underpinned by Vertovian ideas of the camera lens standing in for the biological eyes offering new perspectives on commonplace sights. It has elements of psychogeography and remediates Baudelaire's notion of a flaneur. The participatory and locative media project exploits the assemblages of smartphones and networked technology to pick up the threads left by Vertov's montage techniques to offer contemporary insights into life in our cities where montages can be created from "fleeting moments of existence of the 'I' and the 'non-I' with an immediacy and scale never before possible" (Vickers 2013, 142). Like Schleser, Vickers also uses a map to show the film locations. Vickers found that "social media platforms offer a means for participation, collaboration and distribution or dissemination that was unimaginable even a decade ago" (Vickers 2013, 139) thus extending his practice into new areas engaging the participatory culture phenomena that Jenkins identified in 2008.

Participatory art and digital storytelling projects like these are extending the field of documentary film-making into new areas including participatory and dispersed artworks through mobile media. In these projects, "we see an emplaced visuality that creates and reflects, unique forms of geospatial sociality" (Hjorth and Pink 2014, 54). They manifest the aesthetics of a liquid modernity (Bauman 2003) that requires but a fleeting commitment on a would-be contributor's part. All three projects rely on our ability to document our movements through everyday life and leverage the sociality and emplaced visuality of mobile media. They provide evidence for how participatory media has provided many with agency and has disrupted traditional media distribution power relationships to open up new expressive potentials as we collectively grapple with the everyday realities of networked co-presence, virtual proximity and what these can mean for our everyday social activities and rituals. Without the extreme accessibility of smartphone assemblages these projects would be difficult to execute.

Concluding Comments

At the start of this chapter, I asked—how we better understand practices associated with making films and video with smartphone? If we deploy analytic frames drawn from digital ethnography, we can see that photography and film-making are a way of participating in networked socialities and emplaced visualities. If we take a non-representational approach to film-making and video with smartphones, we can begin to account for the more than representational aspects of content creation. We can observe the Internet's inherent logics of participation, sociability and replicability (Shifman 2007) provide people with agency to generate new visual vernaculars that are in constant flux.

There are normative as well as commercial pressures in social media sites and these influence interactions and content. Participatory media platforms have seeded new forms of interaction through the ability to comment on other people's posts. As Lange (2017) argues, "the fact that people with different media ideologies are encouraged to be interactive within the paradigm of social media will likely to spur discomforting mediated discrepancies for the foreseeable future" (2017, 155). If we juxtapose this feature of social and mobile media with Bauman's notion of a liquid modernity where the fragility, temporariness and predisposition to accelerated rates of change are conditions of contemporary lifeworlds we can begin to see that tensions between giving offence and free speech are not likely to be resolved in the near future.

For creative practice, these tensions provide excellent fodder for extending practices into new forms and new contexts. Hjorth et al. remarked that "As Kester suggests, there is a need to canvas techniques, tools and frameworks for art practice that reimagine the socially engaged dimensions beyond just the "aesthetic play" (2010) in order to shape different forms of inter-subjective affect, identification and agency (2011, 68)" (Hjorth et al. 2017, 289). This applies to smartphone video and film-making where increasingly these creative practices are socially engaged and participate in a relational aesthetics (Bourriaud 2002).

The increasingly popularity of participatory smartphone video projects underscores the participatory nature of mobile media as it shows up within the routines and rituals of everyday life. We can see that smartphones are making their mark by nudging film-making and video into new forms and contexts through new types of curating practices that are emerging and evolving in mobile media ecologies.

Notes

1. http://gerdacammaer.com/?myportfoliotype=mobilarte.
2. https://vimeo.com/181436263?from=outro-embed.

3. https://vimeo.com/168155426.
4. www.now-here-everywhere.org.uk.
5. http://www.24frames24hours.org.nz.
6. http://www.24hours-in.lincoln.ac.uk/.

References

Batty, Craig. 2014. Smartphone Screenwriting: Creativity Technology, and Screenplays-on-the-go. In *Mobile Media Making in an Age of Smartphones*, ed. Berry, Marsha and Schleser Max, 104–114. New York: Palgrave Pivot.

Bauman, Zigmund. 2003. *Liquid Love: On the Frailty of Human Bonds*. Cambridge: Polity Press.

Berkeley, Leo. 2014. Tram Travels: Smartphone Video Production and the Essay Film. In *Mobile Media Making in an Age of Smartphones*, ed. Berry, Marsha and Schleser Max, 25–34. New York: Palgrave Pivot.

Berry, Marsha and Schleser, Max, 2014. Creative Mobile Media: The State of Play. In *Mobile Media Making in an Age of Smartphones* Berry, ed. Marsha and Schleser Max, 1–9. New York: Palgrave Pivot.

Berry, Marsha. 2017. Mobile Filmmaking. In *The Routledge Companion to Digital Ethnography*, ed. Larissa Hjorth, Heather Horst, Anne Galloway, and Genevieve Bell. New York: Routledge.

Bolter, Jay and Grushin, Richard. 1999. *Remediation*. Understanding New Media, Cambridge, Mass: MIT Press.

Bourriaud, Nicolas. 2002. *Relational Aesthetics*, trans. Pleasance, S. and Woods, F. Les presses du reel.

Cardona, Felipe. 2014. Videoloop: A New Edition Form. Special Issue 4, *Journal of Creative Technologies*. https://ctechjournal.aut.ac.nz/paper/videoloop-new-edition-form/.

Cresswell, Tim. 2010. Towards a Politics of Mobility. *Environment and Planning D: Society and Space* 28 (1): 17–31.

Deger, Jennifer. 2017. Curating Digital Resonance. In *The Routledge companion to digital ethnography*, ed. Hjorth, Larissa, Horst, Heather A., Anne Galloway, and Genevieve Bell, 318–328. Taylor & Francis.

Ellis, C.S. and Bochner, A.P. 2006. Analyzing analytic autoethnography: An autopsy. *Journal of Contemporary Ethnography* 35 (4): 429–449.

Farman, Jason. 2014. 'Site-Specifity, Pervasive Computing, and the Reading Interface'. In *The Mobile Story: Narrative Practices with Locative Technology*, ed. Jason Farman. New York: Routledge.

Frankham, B. 2016. Writing with the small, smart screen: Mobile phones, automated editing and holding on to creative agency. *Journal of Writing in Creative Practice* 9 (1–2): 47–66.

Hjorth, Larissa, and Sarah Pink. 2014. New Visualities and the Digital Wayfarer: Reconceptualizing Camera Phone Photography and Locative Media. *Mobile Media & Communication* 2 (1): 40–57.

Hjorth, Larissa, William Balmford, Sharon Greenfield, Amani Naseem, and Tom Penney. 2017. The Art of Play: Ethnography and Playful Interventions with Young People. In *The Routledge Companion to Digital Ethnography*, ed. Larissa Hjorth, Heather Horst, Anne Galloway, and Genevieve Bell. New York: Routledge.

Keep, D., and M. Berry. 2014. Remediating Vertov: Man with a Movie Camera Phone. *Ubiquity: The Journal of Pervasive Media* 2 (1–2): 164–179.

Kelly, Patrick. 2013. North Directed by Patrick Kelly. Australia: Patrick Kelly. MINA 3rd International Mobile Innovation Screening. DVD.

Kelly, Patrick. 2014. Mobile Video Platforms and the Presence of Aura. Special Issue 4, *Journal of Creative Technologies*. https://ctechjournal.aut.ac.nz/paper/mobile-video-platforms-presence-aura/.

Kossoff, Adam. 2014. The Mobile Phone and the Flow of Things. In *Mobile Media Making in an Age of Smartphones*, ed. Marsha Berry and Max Schleser, 35–44. New York: Palgrave Pivot.

Lange, Patricia. 2017. Participatory Complications in Interactive, Video-Sharing Environments. In *The Routledge Companion to Digital Ethnography*, ed. Larissa Hjorth, Heather Horst, Anne Galloway, and Genevieve Bell. New York: Routledge.

Pink, Sarah, and Larissa Hjorth. 2012. Emplaced Cartographies: Reconceptualising Camera Phone Practices in an Age of Locative Media. *Media International Australia* 145: 145–156.

Rieser, M. (ed.). 2011. *The Mobile Audience: Media Art and Mobile Technologies*. Vol. 5. Rodopi.

Schleser, Max. 2014. A Decade of Mobile Moving-image Practice. In *The Routledge Companion to Mobile Media*, ed. Gerard Goggin and Larissa Hjorth. London: Routledge.

Schleser, Max, and Tim Turnbridge. 2013. 24 Frames 24 Hours. *Ubiquity: The Journal of Pervasive Media* 2 (1&2): 205–213.

Schleser, Max R.C, Gavin Wilson, and Dean Keep. 2013. "Small screen and big screen: Mobile film-making in Australasia". *Ubiquity: The Journal of Pervasive Media* 2 (1&2): 118–131.

Shifman, Limor. 2007. Humor in the Age of Digital Reproduction: Continuity and Change in Internet-Based Comic Texts. *International Journal of Communication* 1: 187–209. http://ijoc.org/index.php/ijoc/article/viewFile/11/34.

Vannini, P., 2015. Non-representational ethnography: New ways of animating lifeworlds. *Cultural Geographies* 22 (2): 317–327.

Vickers, Richard. 2013. Mobile Media, Participation Culture and the Digital Vernacular: 24-Hours.in and the Democratization of Documentary. *Ubiquity: The Journal of Pervasive Media* 2 (1&2): 132–145.

CHAPTER 8

Looking over Mobile Media, Creative Practice and Ethnography

A lifeworld, where computers disappear into the fabric of the everyday that was imagined by Mark Weiser (1991) in his famous essay *The Computer for the 21st Century* published in the *Scientific American*, September 1991, has well and truly arrived. In that essay, he dreamed of a day where computers would become akin to writing, which he argues was the first form of information technology. He builds his argument on the premise that technologies quickly become part of the background horizon of everyday life when people have learned how to use them and cites street signs as an example. Ten years ago, the convergence of mobile and social media was in its infancy, and smartphones were quite new. Yet in the second decade of the twenty-first century, networked socialities and co-presence became commonplace features of everyday life. Mobile media ecologies became filled with colliding contexts where time zones and cultural diversity converge. Through social and mobile media, we have unprecedented access to people's intimate thoughts, feelings and creative processes as well as the ability to observe the unfolding of their works in progress through social media networks.

To underscore the degree to which smartphones have become entangled in our lifeworlds, I issued a challenge, in my brief introduction at the very start of the book, to imagine what life would be like if smartphones had not been invented. The mobile phone in its current iteration as the smartphone is an object that has penetrated almost every aspect of our lives. Locative, mobile and social media have converged to form a universal horizon (Husserl 1954) within which we enact creative,

© The Author(s) 2017
M. Berry, *Creating with Mobile Media*,
DOI 10.1007/978-3-319-65316-7_8

communicative and social impulses. In their edited reference volume, *The Routledge Companion to Mobile Media*, Larissa Hjorth and Gerard Goggin (2014) trace the scholarship in mobile media and attest that "with the smartphone heralding a further convergence between locative, social, and mobile media, there is a need for more studies that acknowledge the particular affordances of this convergence" (4). My book also taps this need by paying attention to how the affordances of this convergence have been instrumental in shifting poetry, photography, film-making and video into new zones where these creative practices and processes have become entangled with other social actions and encounters.

Locative, mobile and social media have provided focal points for the research presented in this volume. I have argued that smartphone assemblages have created conditions of possibility and means of production for new and remediated forms of creative expression including creative writing, film-making, photography and video. This convergence has generated new outlets for artists, writers and film-makers through these perpetually emergent media ecologies. The material presented in each chapter provides empirical and material evidence drawn from lived experience through ethnographic and practice-based research where creative artefacts are produced, and practice-led research focusing on new understandings of how the convergence of locative, mobile and social media affects the actions that make up a creative practice. Each chapter focuses on different aspects, action and activities and forms of creative practice to include creative writing, photography and video art, and film-making.

In this final chapter, I first present an overview of what I have discovered about mobile media looking back over the past decade. Then, I look forward seeking new directions about possible futures at the confluences of locative, mobile and social media and creative practice.

Looking Back

When I took the first tentative steps slightly over a decade ago in the direction of mobile phones and mobile media as a subject of enquiry, I had little idea of where my new research direction would lead or how rewarding a decision this would prove to be. At the time, I was teaching undergraduate courses in digital media, and I already had higher research degree candidates who were doing their degrees by creative project and dissertation/exegesis instead of the more traditional thesis. We were exploring nonlinear narrative forms, new forms of archives and narrative databases and experimenting with interactive narrative plotlines in fiction, non-fiction and documentary forms and genres. Social media platforms like Facebook was new and we were still figuring out the different things they could be used for. I recollect that we were also grappling with the quality of video and photographic images for the Internet. Videos would lag, stuttering to create strange visual and audio dissonances and incongruities. Photographs were filled with digital artefacts and glitches. Programmers were racing to develop platforms and browsers that would cope with the data demands of high-resolution images or applications that would be able to compress files without losing too much quality. My own creative practice, at that time, comprised creative writing and photography. My research had, up until this point, used an ethnographic approach.

At the time I commenced my musings and research into mobile media, there was little scholarship written in the area of creative practice research either, and the discussions around the distinctions of practice-led, research-led, practice-based were in their infancy. To briefly recap, practice-led research uses practice and reflects on practice to advance knowledge in a particular practice (Smith and Dean 2009), research-led practice is research where the research leads the practice and practice-based is research that results in the creation of an artefact such as a book of poems, a video art work or a participatory art project.

Justifying practice-as-research was relatively straightforward in digital or new media fields because each project would break new ground—such was the nature of this then largely unchartered terrain. Each development in technology spawned new creative practices as it was appropriated and creatively misused (Farman 2014). Mobile media was no exception. The advances in mobile phones and smartphone assemblages had not yet become "so familiar that they seemingly disappear" but rather they created "a shift to a perspective where we see entirely new ways of using these devices" (Farman 2014, 5). And these new ways of using mobile media affordances provided fingerposts for the research trajectories of several of my higher research degree students as well as for me. In Chap. 1, I presented commonalities and connections between ethnography and creative practice research. My aim was to uncover potential chemistry, synergies and synchronicities because I had an intuition that there is much to be gained from this juxtaposition. My own methodology often was a bricolage of ethnography and creative practice but I had not formally theorized it as such in my previous publications. I was always on the look out for theoretical frames that would allow me to account for the dynamic and slippery aspects of a constantly changing and moving mobile media ecology.

Non-representational theory as explained by Phillip Vannini, an anthropologist, would provide a springboard for my exploration of the potential chemistry between ethnography and creative practice research. Non-representational ethnographic writing techniques inspired by an "ethos of *animation*" (Vannini 2015, 319) can provide an answer to Haseman's call to attend to the performative, inductive and experiential aspects of creative practice research. Furthermore, the fictional turn in ethnographic writing is concerned with aesthetics and poetics as well as with symbolic and sociological structures that govern social activities and everyday practices in specific cultural groups.

Non-representational theory urges us to attend to the more-than-representational aspects so that we can account for embodied, multisensory and background phenomena. For example, in screen production, creative practice research methodology tends to be concerned with verbs rather than nouns—the doing, the making of a film or screenplay rather than what it represents as a cinema studies object. In creative writing, practice-led or practice-based research is more concerned with processes, techniques and craft rather than with exegetical discussions about interpretations of the creative work, which are the stuff of literature and cultural studies. The techniques taken from non-representational theory with their focus on dynamic aspects offer practical as well as evocative ways for creative practitioner scholars to join scholarly conversations about how theory and practice inform each other to contribute and expand knowledge about screen production and creative writing as fields of enquiry.

Experiences, insights, processes and actions from creative practice can be fictionalized in much the same way as those drawn from ethnographic fieldwork, and if this is done with careful attention to providing rich, accurate and thick descriptions to achieve verisimilitude with carefully plotted turning points, a clear sense of what it was like to be there doing the practice may be evoked to inform readers. Photography, video, filmmaking and creative writing are all actions that may reside within a person's creative practice. And these actions may also be integral to the work of an ethnographer. If we inform our methodology with non-representational philosophical epistemologies and ontologies, we can think about the chemistry between ethnography and creative practice research in dynamic ways and push academic writing into exciting new forms and genres.

Mobile and social media are dynamic and participate in lifeworlds that are constantly evolving. Networked media sit on the horizons of our everyday lives and have given rise to new creative vernaculars that are also in constant motion. The appeal of everyday occurrences is manifested on our social media timelines and reflected in much of the material we post. In Chap. 2, I explored how the impulse to share evocative moments has provided fertile ground for the seeds of new visual vernaculars to grow and develop. I conceptualized the assemblages of locative, mobile and social media and their relationship to physical locations and the offline as zones of entanglement where a sharp distinction between online and offline is no longer useful.

New vernaculars have emerged and have found expression in participatory cultures in the guise of numerous interest groups focussing on aspects of creative writing and photography that have flourished in social media. Through such networked interest groups, people seek to build up an audience and we see visual vernaculars such as emoticons and memes being used to build up a rapport with others that may translate into the audience and a regular following for people's creative outputs. Socialities have become entangled with creative practices. These socialities have found expression through emplaced visualities (Pink and Hjorth 2012) and these in turn have manifested in the sharing of evocative moments such as photographs of sunsets or shadowy reflective streetscapes. If one has smartphone camera apps, one can take an ordinary image of a streetscape and transform it into an atmospheric one with a few taps on the smartphone screen.

I found that the combination of ethnography and creative practice research framed by non-representational theory in the semblance of Ingold's notion of a zone of entanglement really helped me gain insights into how new creative vernaculars come into being. His conceptualization of a zone of entanglement exposes a pathway that meanders its way through commonly perceived boundaries such as geographic places and distances, and binaries such as online and offline to uncover how creative expressions and forms circulate and come into being as a creative visual vernacular practices within zones of entanglement generated by smartphone assemblages and mobile media.

The kinds of shapes mobile media will assume in the future are difficult to predict—perhaps augmented or a hybrid mixed reality will become the norm given the success of *Pokemon Go* in 2016, perhaps virtual reality applications will become ubiquitous and invisible following the trajectory Weiser envisaged for personal computing back in 1991, perhaps all our smartphones will have VR capabilities, or perhaps none these will come to pass. Nevertheless, we can be reasonably sure that creative vernaculars will continue to evolve and visual media will continue to be more than representational and will continue to inspire evocative expressions as people continue to play in mobile media spaces and enact an emplaced visuality.

> Kim walked in the shopping district – there were advertisements in footpath screens embedded in the pavements. Sometimes guerrilla artists took them over posting video art and images that looked like graffiti from 2015. Today though it was just the usual dross – special two for one offers from fast food chains and free coffees from Ubiquitous Coffee if you shared their ad using Instagram to your account. He looked up instead and saw a pattern of tiles that seemed to be repeated on the walls of various buildings. His curiosity was piqued. The patterns looked like they could be scanned like the old QR codes. He held his smartphone camera up to scan the tiles and the pattern dissolved into a 19th century streetscape much like Google street view using 360 degrees and a story made up of text scrolling over the streetscape about a man who had lived here and ran a bookshop in this exact spot where now there was an apartment block of twenty stories. There was audio as well and you could select a translation of the story from twenty different languages or you could listen to it in the original. There was also an option to watch it in full VR if you had the right headset. The story was one of romance, loss and longing – the bookshop owner's bride to be died in an influenza epidemic and so he never married. It was said that he died in the shop and that many believed that he and his bride to be still haunted the site where the bookshop once stood on nights when the moon was full. Intrigued, Kim looked for the next pattern piece to scan.

Selfies are an example of a new creative visual vernacular and are often used to perform emplaced visuality. In part, they are a remediation of self-portraits and postcards and they seem to figure prominently in social media. They are networked images and are a relatively recent cultural phenomenon. They participate in what Bauman (2003) describes as a liquid modernity and the conventions and cultural norms for what is deemed socially acceptable are hotly contested in mainstream media. Indeed, many popular media commentators point to selfies as evidence of contemporary social ills and a perceived rise of narcissistic selfishness (see, e.g. Hinde 2016). A counter to such ruminations is provided by academic scholarship that investigates the reasons why people post. These studies provide broader contexts for the special place of selfies within other everyday practices with social media. Miller et al. (2016) undertook a landmark multi-sited global study into the reasons why people post and discovered that selfies reflect local cultural norms and that each location has its own nuances. On the other hand, research such as Rettburg's study (2014) use representational approaches to situate selfies within a genealogy of other traditions of self-examination and self-representation like autobiography and self-portraiture.

My own approach to the selfie phenomenon was ethnographic informed by non-representational theory. I sought to uncover some of the intentions and functions of selfies. I posed a question about what we wanted to see when we posted selfies. The functions I adduced from my ethnographic observations and field notes included postcards, points of contact, showing off, self-affirmations (self-expression), self-promotion, self-documentation, self-invention and reinvention. The forms of selfies I discovered included images of people's feet and their shadows as well as faces. This parallels Miller et al. (2016) findings of "footies" being used as a form of selfie in Chile and Italy. Selfies have the capacity to reveal both similarities and differences in various cultural groups with regard to what is regarded as socially acceptable and will continue to provide conceptual artists with inspiration. They have evolved into a creative visual vernacular with local inflections and nuances.

In Chaps. 2 and 3, I focused on how locative, mobile and social media assemblages have initiated the emergence of new creative visual vernaculars. In Chap. 4, I returned to a fundamental concept drawn from classical ethnography—"being there". I put forward a proposition that the extreme accessibility of smartphones proffers a path-breaking opportunity to summon a sense of place. I used a rubric of three key concepts to digital ethnography (Hjorth and Pink 2014)—mobility, digital co-presence and digital wayfaring as the basis for a braided account of mobile art-making practices such as shooting photographs and video that seek to evoke the atmospheric and corporeal aspects of being in particular place at a particular time. The actions comprising photography and video by their very nature are embodied. Smartphones afford the opportunity to edit images in the camera and to share these with digitally co-present others (synchronously and asynchronously). I discovered that the everyday and the evocative constantly intersect in mobile media lifeworlds and have shifted the relational aesthetics of screen production and photography into new forms through the new kinds of visualities identified by Pink and Hjorth (2012).

From layering together an account of some of the complexities of conveying a sense of what it is like to be there, in a specific place, at a specific time and experiencing specific sensations, feelings and atmospheres with smartphone assemblages in Chap. 4, in Chap. 5, I turned my attention to revisiting the assemblages of locative, mobile and social media as generators of virtual spaces where creative writers may gather to improvise and collaborate creatively. I first became involved in online poetry networks in 2010. In this chapter, I recount my own lived experience of these inclusive virtual spaces that transcend national and cultural boundaries. These spaces continue to thrive and expand as outlets for creative writers and artists. These spaces also allow creative writers and artists to form relationships, and to play and perform together and for each other in virtual playgrounds, studios and writing groups. The ability to improvise and collaborate with other poets is dependent on relational aesthetics (Bourriaud 2002)—a shared understanding of how forms work—as well as on digital co-presence.

Poetic and evocative expressions inspired by place and landscape can instigate new ways of thinking about the places we inhabit. I discuss these in Chap. 6 through an ongoing research project—*Poetry4U*. Locative mobile media present openings for the creation of new cultural spaces that may be curated or not. The *Poetry4U* website is an ongoing project that pins poetry to place. The project is an evolution of earlier research where a colleague from computer science, and I explored the potential of Bluetooth servers as a way of seeding locative media within specific places (Berry and Hamilton 2010). The technology continues to develop at a breathtaking pace where no sooner than we get used to how something works, it shifts into its next iteration. The potential of augmented reality for locative media and place-specific storytelling is immense (Farman 2014), and we are only just beginning to see this. Currently, the creation of augmented reality apps does require specialized knowledge, but it is quite possible that in the future, the creation of augmented reality will be as simple as applying a filter to a photograph or video shot with a smartphone. This will open up new possibilities as creative writers, film-makers and artists immerse themselves in the potential of augmented reality. New cultural spaces and forms will emerge as a consequence. We can only imagine what these will be like.

Jess prowls in the laneway at dusk. The rain has stopped. The sky is a deep violet blue and the street lights a golden compliment oozing onto the cobbles. Jess stops, breathless. An ah-ha moment – an arrival of sorts. She inhales deeply, takes her smartphone from her coat pocket as she exhales luxuriantly. She shoots a short clip, tags it with location and overlays a haiku about what caught her breath. She posts it. She decides to make a film using haiku in the form of walking trail – but for now this will be an improvised performance. She receives several responses in kind – others have joined in the improvisation and are writing haiku in alleys. What began, as a solo performance has become collaboration? This is a fascinating twist.

In Chap. 7, I explored mobile film-making, photography and video through the lens of creative arts practice research as well as ethnography. Cameras like computers have disappeared into the background horizons of our everyday life. In the pre-digital age, we would need to remember to take our cameras and would often only take these with us on special occasions or pre-planned events or shoots. We would have to carefully frame images so as not to waste valuable film. If we were using black and white stock, we would need to be vigilant about hue values and contrast, about shadow and light. We would also kick ourselves for not having a camera if a particular sight or event stirred our muses.

The days when shooting images—still and moving—required such special efforts have long since passed. The days when the film was processed in darkrooms and specialized laboratories are now the focus of nostalgia. We can see what we have captured straight away which was not the case in the pre-digital age of photography. We can share the outputs of our creative impulses almost instantaneously using our smartphones. The affordances and assemblages of smartphones have expanded the field of film-making and video into participatory art projects that take good advantage of Internet's inherent qualities of participation, sociability and replicability (Shifman 2007). The rise of spreadable media where distribution is no longer firmly controlled by corporations (Jenkins et al. 2013) has rattled traditional media industrial complexes. All of these advances are hallmarks of locative, mobile and social media convergence in the second decade of the twenty-first century.

Looking Forward

There is always a tension between looking back through the known and imagining futures, and the future imaginings are subject to what apps and affordances will capture the popular imagination. Although cameras have slipped into the backgrounds of our lifeworlds, the potential for their use in academic research is still emerging. Smartphones, with their video capabilities, are an incredibly useful tool for social research as well as for creative practice. Sarah Pink has been an advocate of video as a method for over a decade with her book Doing Visual Ethnography, which is now in its third edition. Video as a method for social research is a growing area of academic activity according to Anne Harris (2016). Harris (2016) reveals the extent to which video has become important: "The role of video as method continues to expand in almost every discipline" (2016, 4). She also notes that it is easy to become overwhelmed. Harris argues that video is both method and methodology and its "overall evolution as a research approach continues to expand the possibilities for knowledge creation—but, perhaps more importantly, it expands the possibilities for real impacts in the communities that it seeks to affect and reflect" (Harris 2016, 14).

In Chap. 1, I explored how ethnographic approaches with smartphones and other mobile devices can extend creative practice research into new directions. To extrapolate on Harris's position as video being both method and methodology, I would suggest we are only just beginning to understand the potentials of smartphones as a tool for social research as well as for creative practice. If we conceive of the things we can do with smartphones as both method and methodology, we can also expand the possibilities for knowledge creation through creative practices and to seek out possible pathways of the impact of these in our lifeworlds.

The confluence of locative, mobile and social media has restless fluidity. New creative potentials have emerged as we collectively grapple with the everyday realities of networked co-presence and virtual proximity. There is always oscillation between the known and the unknown, the new and the remediated. Smartphone camera practices are vernacular, expressive and often have a dimension of performance. They are also multi-layered. The opportunities for artists, film-makers and writers to improvise with mobile media assemblages and affordances will continue to expand at an ever-increasing rate. The trick will be to keep up yet Bauman's philosophical concept of a liquid modernity suggests that adaptation to new things is as swift as their development.

Mixed reality experiences including ambient games, virtual and augmented reality have changed our perceptions of place irrevocably. They have also had a profound effect on our engagement with place. Walking on any urban sidewalk now involves dodging pedestrians distracted by their screens. One only needs to look at the bright yet ephemeral flash of Pokemon Go in mid-2016 and associated concerns about pedestrian safety (Pietrowicz 2016) to begin to imagine how the convergence of mobile, locative and social media might manifest in the future and to see how difficult it is to predict the success or otherwise of specific technological developments. Google Glass threatened to disrupt our lifeworld in 2012 but failed due to a public outcry about privacy concerns (Bilton 2015). What mobile media assemblages can mean for our everyday social activities and rituals is a dynamic cluster where relations are constantly forming, reconfiguring and transfiguring the fabric of our lifeworlds.

Go-Pro and dashboard cameras are commonplace and perhaps it is only a matter of time until they have the seamless networked capability of smartphones. On the other hand, VR headsets such as Google cardboard have become affordable, but their uptake seems to be dependent on the availability of content and has not been as rapid as the industry pundits predicted. Indeed, as Watson concludes in her comprehensive report on VR and 360 applications and potentialities for news:

> Early ethnographic research by the BBC suggests we have a long way to go before 'public service' content (which includes news) and VR headsets will fit into people's lives. Discovery on VR platforms is poor, with little effective curation – users need to be able to find content that fits their needs better. The research concludes that the public are confused about VR. (Watson 2016, n.p.)

Curation of content seems to be the key to success to VR, and I suggest it is also important to the success of future developments and directions in mobile media. However, we can also apply knowledge gained through the popularity of smartphone camera apps and how this has encouraged people to engage in creative practices with photography and video to VR, 360 applications and AR (augmented reality).

In 2010, I wrote that locative and mobile media utilizing "novel combinations of ubiquitous technology can flourish in the gaps helping to create 'new shadows and opacities'" (Berry and Hamilton 2010, 53), to augment places with narrative overlays. At the time I wrote those words, locative media applications and technologies were very basic. Locative media installations tended to rely on localized technology such as Bluetooth. The ease with which we will be able to aggregate and play with narratives connected with the place will continue to improve and to uncover the depths and layers of the complex relations we have with place.

Hjorth et al. (2016), in a recent volume documenting and theorizing mobile media art in the Asia-pacific region, refer to Mimi Sheller's (2014) position that mobile media art is much more than simply screening digital art on smartphones and tablets. Sheller, in her riposte to Claire Bishop's claim that contemporary digital art was "simply about screening digital art on portable devices" stated that mobile art "has in fact expanded the spatial and social field in which art takes place by experimenting with the mobile interface as a bridge between digital and physical space, a hybrid mediation of human sensory perception and technological connectivity" (Sheller cited in Hjorth et al. 2016, 179). My own research—creative practice and ethnography—has explored the bridge between the digital and the physical, and the new ecologies mobile media has provided for creative practices—both my own and those of my fellow travellers.

Mobile media assemblages have become more and more seamless and the affordances are increasingly more sophisticated. We have reasonably easy access to landscape perspectives hitherto difficult to achieve because of cameras attached to drones. Film festivals focussed on drone film-making are becoming more popular and according to drone filmmaker, Jay Worley (Airvus 2017), aerial shots are taking the art of film-making into fresh areas that are shifting film-making aesthetics. It is easy to imagine drone cameras that interface with GPS and Wi-fi to provide live and multiple perspectives and points of view pinned to places (Airvus 2017). Mobile media assemblages will endure to nourish and shape how stories can be told as well as which stories can be told. Smartphones will still challenge the lines between online and offline, material and immaterial. They will continue to signpost new kinds of creative activities that remediate old forms and point the way to new forms and practices.

A non-representational theory may yet provide even more frames for creative practice research. These frames may extend and cross-pollinate across both methods and methodologies in both digital ethnographic and creative practice research. Any predictions for mobile media futures are fraught and uncertain; however, I feel that it is reasonable to predict that creative practitioners will tarry to play in the convergence between locative, mobile and social media. New creative vernaculars will flourish, nourished by the impulses and interplays between artists, writers, film-makers and photographers. In short, to conclude, new forms and genres of mobile media art will continue to evolve through creative actions and practices in response to hybrid screen ecologies that bridge the physical and the digital, the material and the immaterial, the tangible and the intangible.

REFERENCES

Airvuz. 2017. The Drone Dish: Jay Worsley [Video]. <https://www.airvuz.com/video/The-Drone-Dish-Jay-Worsley?id=58dd5517cdad787c3ac6e1ad>. Accessed 5 May 2017.

Bauman, Zigmund. 2003. *Liquid Love: On the Frailty of Human Bonds.* Cambridge: Polity Press.

Berry, Marsha, and Margaret Hamilton. 2010. Mobile Computing Applications: Bluetooth for Local Voices. *Journal of Urban Technology* 17 (2): 37–55. doi:10.1080/10630732.2010.515084.

Bilton, Nick. 2015. Why Google Glass Broke. https://www.nytimes.com/2015/02/05/style/why-google-glass-broke.html?_r=0.

Bourriaud, N. 2002. Relational Aesthetics, trans. S. Pleasance and F. Woods. Les presses du reel.

Farman, Jason. 2014. 'Site-Specifity, Pervasive Computing, and the Reading Interface'. In *The Mobile Story: Narrative Practices with Locative Technology*, ed. Jason Farman, New York: Routledge.

Harris Anne. 2016. *Video as Method.* Oxford: Oxford University Press.

Hinde, Natasha. 2016. Jennifer Saunders: Girls' Quest For the Perfect Selfie Is Making Them 'Ill'. *Huffington Post 17/09/2016.* http://www.huffingtonpost.com.au/2016/11/28/jennifer-saunders-believes-girls-quest-for-the-perfect-selfie-is-making-them-ill/?utm_hp_ref=au-selfies.

Hjorth, Larissa, and Gerard Goggin. 2014. *The Routledge Companion to Mobile Media.* New York: Routledge.

Hjorth, Larissa, and Sarah Pink. 2014. New Visualities and the Digital Wayfarer: Reconceptualizing Camera Phone Photography and Locative Media. *Mobile Media & Communication* 2 (1): 40–57.

Hjorth, Larissa, Sarah Pink, Kristen Sharp, and Linda Williams. 2016. *Screen Ecologies: Art, Media, and the Environment in the Asia-Pacific Region.* Cambridge: MIT Press.

Husserl, Edmund. 1954 (1970). *The Crisis of European Sciences and Transcendental Phenomenology.* Evanston: Northwestern University Press. http://s3.amazonaws.com/academia.edu.documents/34337886/

Husserlcrisis.pdf?AWSAccessKeyId=AKIAIWOWYYGZ2Y53UL3A&Expire
s=1488518588&Signature=q4ETJQ95ohYGnfSm8wHs%2BFXGctU%3D
&response-content-disposition=inline%3B%20filename%3DThe_Crisis_of_
European_Sciences_and_Tran.pdf.

Jenkins, Henry, Sam Ford, and Joshua Green. 2013. *Spreadable Media: Creating Value and Meaning in a Networked Culture.* New York: NYU press.

Mark, Weiser. 1991. The Computer for the 21st Century. *Scientific American.* https://www.ics.uci.edu/~corps/phaseii/Weiser-Computer21stCentury-SciAm.pdf.

Miller, Daniel, Costa, Elisabetta, Haynes, Nell, McDonald, Tom, Nicolescu, Razvan, Sinanan, Jolynna, et al. 2016. *How the World Changed Social Media.* UCL Press. pdf available ucl.ac.uk/ucl-press.

Pink, S., and L. Hjorth. 2012. Emplaced Cartographies: Reconceptualising Camera Phone Practices In an Age of Locative Media. *Media International Australia* 145: 145–156.

Pietrowisz, Stacey. 2016. Distracted Walking—Pokemon Go Causing Accidents and Injuries. https://www.bostoninjurylawyer-blog.com/2016/07/distracted-walking-pokemon-go-causing-injuries-accidents.html.

Rettburg, Jill Walker. 2014. *Seeing Ourselves Through Technology: How We Use Selfies, Blogs and Wearable Devices to See and Shape Ourselves.* New York: Palgrave Macmillan.

Shifman, Limor. 2007. Humor in the Age of Digital Reproduction: Continuity and Change in Internet-Based Comic Texts. *International Journal of Communication.* 1 (2007): 187–209. http://ijoc.org/index.php/ijoc/article/viewFile/11/34.

Smith, Hazel, and Roger Dean. (eds.) 2009. *Practice-led Research, research-led Practice in the Creative Arts.* Edinburgh University Press.

Vannini, Phillip. 2015. Non-representational Ethnography: New Ways of Animating Lifeworlds. *Cultural Geographies* 22 (2): 317–327.

Watson, Zillah. 2016. VR for News: The New Reality? Digital News Publications: Reuters Institute for the Study of Journalism. Oxford University. http://digitalnewsreport.org/publications/2017/vr-news-new-reality/#conclusion.

Index

The manufacturer's authorised representative in the EU is Springer
Nature Customer Service Centre GmbH, Europaplatz 3, 69115 Heidelberg,
Germany. If you have any concerns regarding our products, please
contact ProductSafety@springernature.com

Printed and bound by CPI Group (UK) Ltd, Croydon, CR0 4YY
27/04/2026
02097629-0001